# Lecture Notes in Physics

## The Editorial Policy for Proceedings

The series Lecture Notes in Physics reports new developments in physical research and teaching – quickly, informally, and at a high level. The proceedings to be considered for publication in this series should be limited to only a few areas of research, and these should be closely related to each other. The contributions should be of a high standard and should avoid lengthy redraftings of papers already published or about to be published elsewhere. As a whole, the proceedings should aim for a balanced presentation of the theme of the conference including a description of the techniques used and enough motivation for a broad readership. It should not be assumed that the published proceedings must reflect the conference in its entirety. (A listing or abstracts of papers presented at the meeting but not included in the proceedings could be added as an appendix.)

When applying for publication in the series Lecture Notes in Physics the volume's editor(s) should submit sufficient material to enable the series editors and their referees to make a fairly accurate evaluation (e.g. a complete list of speakers and titles of papers to be presented and abstracts). If, based on this information, the proceedings are (tentatively) accepted, the volume's editor(s), whose name(s) will appear on the title pages, should select the papers suitable for publication and have them refereed (as for a journal) when appropriate. As a rule discussions will not be accepted. The series editors and Springer-Verlag will normally not interfere with the detailed editing except in fairly obvious cases or on technical matters.

Final acceptance is expressed by the series editor in charge, in consultation with Springer-Verlag only after receiving the complete manuscript. It might help to send a copy of the authors' manuscripts in advance to the editor in charge to discuss possible revisions with him. As a general rule, the series editor will confirm his tentative acceptance if the final manuscript corresponds to the original concept discussed, if the quality of the contribution meets the requirements of the series, and if the final size of the manuscript does not greatly exceed the number of pages originally agreed upon.

The manuscript should be forwarded to Springer-Verlag shortly after the meeting. In cases of extreme delay (more than six months after the conference) the series editors will check once more the timeliness of the papers. Therefore, the volume's editor(s) should establish strict deadlines, or collect the articles during the conference and have them revised on the spot. If a delay is unavoidable, one should encourage the authors to update their contributions if appropriate. The editors of proceedings are strongly advised to inform contributors about these points at an early stage.

The final manuscript should contain a table of contents and an informative introduction accessible also to readers not particularly familiar with the topic of the conference. The contributions should be in English. The volume's editor(s) should check the contributions for the correct use of language. At Springer-Verlag only the prefaces will be checked by a copy-editor for language and style. Grave linguistic or technical shortcomings may lead to the rejection of contributions by the series editors.

A conference report should not exceed a total of 500 pages. Keeping the size within this bound should be achieved by a stricter selection of articles and not by imposing an upper limit to the length of the individual papers.

Editors receive jointly 30 complimentary copies of their book. They are entitled to purchase further copies of their book at a reduced rate. As a rule no reprints of individual contributions can be supplied. No royalty is paid on Lecture Notes in Physics volumes. Commitment to publish is made by letter of interest rather than by signing a formal contract. Springer-Verlag secures the copyright for each volume.

## The Production Process

The books are hardbound, and the publisher will select quality paper appropriate to the needs of the author(s). Publication time is about ten weeks. More than twenty years of experience guarantee authors the best possible service. To reach the goal of rapid publication at a low price the technique of photographic reproduction from a camera-ready manuscript was chosen. This process shifts the main responsibility for the technical quality considerably from the publisher to the authors. We therefore urge all authors and editors of proceedings to observe very carefully the essentials for the preparation of camera-ready manuscripts, which we will supply on request. This applies especially to the quality of figures and halftones submitted for publication. In addition, it might be useful to look at some of the volumes already published. As a special service, we offer free of charge LATEX and TEX macro packages to format the text according to Springer-Verlag's quality requirements. We strongly recommend that you make use of this offer, since the result will be a book of considerably improved technical quality. To avoid mistakes and time-consuming correspondence during the production period the conference editors should request special instructions from the publisher well before the beginning of the conference. Manuscripts not meeting the technical standard of the series will have to be returned for improvement.

For further information please contact Springer-Verlag, Physics Editorial Department V, Tiergartenstrasse 17, W-6900 Heidelberg, FRG

Peter L. Smith  Wolfgang L. Wiese (Eds.)

# Atomic and Molecular Data for Space Astronomy

## Needs, Analysis, and Availability

A Selection of Papers Presented at
the Joint Commission Meeting III
of the 21st IAU General Assembly
Held in Buenos Aires, Argentina
23 July – 1 August 1991

Springer-Verlag
Berlin Heidelberg New York
London Paris Tokyo
Hong Kong Barcelona
Budapest

**Editors**

Peter L. Smith
Atomic and Molecular Physics Laboratory
Harvard College Observatory
Harvard-Smithsonian Center for Astrophysics
60 Garden Street, Cambridge, MA 02138, USA

Wolfgang L. Wiese
Atomic Physics Division
National Institute of Standards and Technology
Gaithersburg, MD 20899, USA



ISBN 0-387-97909-3 Springer-Verlag New York Berlin Heidelberg
ISBN 3-540-97909-3 Springer-Verlag Berlin Heidelberg New York


Printed in Germany

Typesetting: Camera ready by author/editor
58/3140-543210 - Printed on acid-free paper

# Preface

In recent years, as new astronomical satellite spectrometers have been developed and planned, questions about the atomic and molecular data base have arisen. In particular, astronomers have asked: Are the atomic and molecular data required for space astronomy available and of adequate quality, and, if not, what improvements are necessary?

In order to explore these questions, Commission 14 (Atomic and Molecular Data) of the International Astronomical Union (IAU) proposed that a Joint Commission Meeting on the subject of *Atomic & Molecular Data for Space Astronomy: Needs and Availability* be held at the XXI$^{st}$ General Assembly of the IAU. The proposal was supported by Commissions 10 (Solar Activity), 12 (Solar Radiation and Structure), 15 (Physical Study of Comets, Minor Planets, and Meteorites), 16 (Physical Study of Planets), 29 (Stellar Spectra), 34 (Interstellar Matter), 35 (Stellar Constitution), 36 (Theory of Stellar Atmospheres), and 44 (Astronomy from Space).

This book comprises written versions of the majority of the presentations made at the meeting, which was held on 26 July, 1991, in Buenos Aires, Argentina. Abstracts of all the presentations can be found in *Highlights of Astronomy, Vol. 9*, edited by J. Bergeron and published by Kluwer Academic Publishers. We hope that this book will be valuable to laboratory astrophysicists who are trying to provide the data needed for analysis of data from astronomical spectroscopic satellites.

The editors thank the IAU for graciously permitting publication of these proceedings in this format, and the Presidents and members of the IAU Commissions, who shared in recommending and supporting this Meeting. The editors also thank the members of the organizing committee, S. Sahal-Brechot, Past President of Commission 14 (Atomic & Molecular Data), M. C. E. Huber, and D. C. Morton, for their advice and assistance in planning and running the meeting, and we thank the authors for their diligent efforts at succinctly summarizing their oral presentations.

The editing of this volume was supported in part by NASA Grants NAGW-1596 to Harvard University and W-15331 to the U.S. National Institute of Standards and Technology. A U.S. National Science Foundation travel grant provided partial support for travel to the IAU meeting.

# Table of Contents

# 1

# Atomic and Molecular Data for Space Astronomy: Needs and Availability

**Peter L. Smith[1] and Wolfgang L. Wiese[2]**

ABSTRACT Atomic and molecular spectroscopic data are essential for the analysis of astronomical spectra and for modelling conditions in astronomical objects where photon-induced processes are important. In order to assess the atomic and molecular data required for analyzing spectroscopic observations from existing and planned orbiting observatories, and to provide space astronomers with a current inventory of atomic and molecular data compilations, a meeting was held as part of the XXI$^{st}$ General Assembly of the International Astronomical Union in Buenos Aires, Argentina, in July of 1991. This book comprises written versions of the majority of the presentations made at the meeting. This chapter reviews, in general terms, the use of atomic and molecular spectroscopic data in astrophysics and discusses current state of laboratory capabilities for obtaining these data.

## 1.1 Introduction

As new, increasingly powerful, astronomical satellite spectrometers have been developed and planned, concomitant concerns about the atomic and molecular data base have arisen: Are the atomic and molecular data required for space astronomy available and of adequate quality, and, if not, what improvements are necessary?

In order to explore these questions, Commission 14 (Atomic and Molecular Data) of the International Astronomical Union (IAU) proposed that a Joint Commission Meeting (JCM) on the subject of *Atomic & Molecular Data for Space Astronomy: Needs and Availability* be held at the XXI$^{st}$ General Assembly of the IAU in Buenos Aires, Argentina, in July of 1991. The motivations were, *(i)*, to review on-going and planned astronomical

[1]Harvard-Smithsonian Center for Astrophysics, Cambridge, MA, 02138, USA

[2]National Institute of Standards and Technology, Gaithersburg, MD, 20899, USA

spectroscopy from space and the atomic and molecular data required to support such activities, and, *(ii)*, to provide space astronomers with a current inventory of atomic and molecular data compilations and new data-generation activities, the Opacity Project in particular. The subject matter of the JCM covered atomic and molecular spectroscopic data for the extreme ultraviolet to submillimeter wavelength range (i.e., from ~10 nm to ~1 mm).

The proposal was supported by Commissions 10 (Solar Activity), 12 (Solar Radiation and Structure), 15 (Physical Study of Comets, Minor Planets, and Meteorites), 16 (Physical Study of Planets), 29 (Stellar Spectra), 34 (Interstellar Matter), 35 (Stellar Constitution), 36 (Theory of Stellar Atmospheres), and 44 (Astronomy from Space) of the IAU. The large number of sponsoring commissions, significantly greater than for other JCMs, testifies to the breadth of support for improvements on the spectroscopic database for atoms and molecules that are directly observed, or that are components of models in contemporary astrophysical research.

Twelve presentations, by an international group of astronomers, physicists, and chemists involved in astronomical spectroscopy or laboratory astrophysics,[3] were made at the JCM. Seven of the authors have contributed to this summary volume. Abstracts of all the presentations can be found in *Highlights of Astronomy, Vol. 9*, edited by J. Bergeron [Kluwer Academic Publishers]. An eighth paper was explicitly solicited for this book: W. H. Parkinson complements the contribution of W. C. Martin (*Sources of Atomic Spectroscopic Data for Astrophysics*) by providing a *Summary of Current Molecular Databases.*

## 1.2 Atomic and Molecular Data in Astrophysics

Because almost all of our knowledge of the Universe reaches us in the form of photons, atomic and molecular spectroscopic data are essential components of research in astronomy and astrophysics. The accuracy with which physical conditions in objects studied by astronomers can be inferred from spectroscopic observations depends directly on the breadth and precision of the data available for atomic and molecular processes. In addition, reliable atomic and molecular spectroscopic data are required for models of photon-driven physical and chemical processes and for models of the steady state properties of astronomical objects that absorb or emit photons. The increasing sophistication, precision, and range of observational techniques and theoretical models continually create new demands for more and better data on atomic and molecular properties.

---

[3] The term 'laboratory astrophysics' is intended to include both measurement *and calculation* of atomic and molecular spectroscopic parameters for astronomy.

There has been a long history of intimate connections between astrophysics and atomic and molecular spectroscopy. Prior to 1962, The Astrophysical Journal was explicitly subtitled *An International Review of Spectroscopy and Astronomical Physics.* As astronomical spectroscopy extends further its frontiers in wavelength and sensitivity, the futures of astronomy and of atomic and molecular spectroscopy will continue to be closely linked. A large number of articles in such journals as the Journal of Chemical Physics, Chemical Physics, Chemical Physics Letters, the Journal of Geophysical Research, the Journal of Molecular Spectroscopy, etc., cite astronomical applications as stimuli and justifications for the investigations that they report.

## 1.3 Astronomical Spectroscopy from Space

The 1990's will see the deployment of powerful new satellite instruments for astronomical spectroscopy and unprecedented concomitant growth in the quality and variety of astronomical spectroscopic data. The capabilities of some of these satellite spectrometers are reviewed in the chapters that follow.

Because space astrophysics missions are conspicuous engineering efforts, instrument signals are sometimes equated with scientific success. However, critical attention to the scientific data is needed. Robust experimental and theoretical programs in laboratory astrophysics are required to optimize mission planning and, ultimately, to transform returned spectroscopic data into scientific knowledge through calibration, analysis, and interpretation.

## 1.4 Laboratory Astrophysics

Quantitative analyses of the spectra of astronomical sources, and of the processes that populate the atomic and molecular energy levels that give rise to emission and absorption by these sources, require accurate data on transition frequencies (or energies) and probabilities, electron impact excitation, deactivation, and ionization cross sections, photoionization and photodetachment cross sections, radiative and dielectronic recombination and radiative attachment rate coefficients, and cross sections for heavy particle collisions involving charge transfer, excitation, and ionization. If molecules are present, processes such as radiative association, rotational and vibrational excitation, ion-molecule and neutral-particle chemical reactions, dissociative recombination, photodissociation, and collision-induced absorption must be quantitative described. Many of these data are, at present, unavailable: For example, even for the most simple and fundamental molecule, $H_2$, there are uncertainties regarding the cross sections for collisional excitation of the vibration-rotation levels from which we observe emission.

However, the laboratory research, including atomic and molecular theory, that yields the data required for astronomical spectroscopy has not been very active during the past decade, and the number of atomic and molecular physicists and chemists responding to the needs of astronomy has declined considerably in the absence of continuing strong support for laboratory astrophysics.

Internal developments in atomic and molecular physics have aggravated this situation: In the past, much of the fundamental research in atomic and molecular physics and chemistry has provided spectroscopic data needed by astronomers. But laboratory astrophysics, is, for the most part, no longer at the forefront of basic research in physics and chemistry, where research is now driven by more fundamental considerations such as: quantum suppression of chaos, scattering at ultra-low temperatures, multiphoton spectroscopy, above threshold ionization and dissociation, intense field effects, planetary atoms, cavity effects in quantum electrodynamics, laser-induced energy transfer, multichannel time-dependent photoionization and photodissociation, and femtosecond spectroscopy – topics whose connection to astronomy is ultimately profound, but which will only incidentally produce data useful to astronomers.

Because the problems raised by astronomers are no longer at the cutting edge of research in atomic and molecular physics and chemistry, the astronomy community will have to provide support for laboratory astrophysics research in a more systematic way and with a longer term perspective than it has in the past.

The five reviews on data needs for space astronomy presented in these proceedings highlight the principal features of major, current and future, satellite spectroscopic observations and point to shortcomings in the atomic and molecular database. At the present slow pace of activity in atomic and molecular laboratory astrophysics, it is impossible that these needs can be met in the foreseeable future at the quality levels desired. In fact, it is much more likely that the discrepancies between needs and available data will widen drastically as many new astronomical data are obtained from the new generation of satellite observatories.

## 1.5 Summary

The first group of papers in this volume reviews the capabilities of the many spectroscopic astronomical satellites that have obtained, or will soon produce, data. The final three papers provide a current inventory of atomic and molecular spectroscopic data compilations.

Accurate atomic and molecular spectroscopic data are essential for the analysis of such spectra and for modelling conditions in astronomical objects where photon-induced processes are important. The authors provide representative examples, at wavelengths from the extreme ultraviolet to the submillimeter, of shortcomings in the current database, shortcomings that will continue to increase quickly and dramatically during the next decade as ambitious space missions are launched while laboratory astrophysics capabilities are expected to languish.

The data requirements outlined are diverse: New, high-quality data are required for many chemical elements and stages of ionization, and many existing data need to be significantly improved in order to provide the margins of security in the results that are required to draw definite conclusions.

## 1.6 Acknowledgements

The authors thank the IAU, for graciously permitting publication of these proceedings in this format, and the Presidents and members of the IAU Commissions, who shared in recommending and supporting this Joint Commission Meeting at the XXI$^{th}$ General Assembly of the IAU.

The authors also thank the members of the organizing committee, S. Sahal-Brechot, Past President of Commission 14 (Atomic & Molecular Data), M. C. E. Huber, and D. C. Morton, for their advice and assistance in planning and running the meeting, and, of course, we thank the authors for their diligent efforts at succinctly summarizing their oral presentations.

We also thank a number of astronomers, too numerous to list, who encouraged us in our endeavors at increasing the awareness of the inadequacy of the existing database in atomic and molecular spectroscopic data for astronomy. Many of the thoughts expressed in this introduction originated with these colleagues, especially A. Dalgarno.

PLS particularly thanks W. H. Parkinson for patiently allowing important laboratory measurements to be postponed while the JCM was organized and while these proceedings were edited.

This work was supported in part by NASA Grants NAGW-1596 to Harvard University and W-15331 to the U.S. National Institute of Standards and Technology. A U.S. National Science Foundation travel grant provided partial support for travel to the IAU meeting.

# 2

# The Hubble Space Telescope – Scientific Problems and Laboratory Data Needs

Steven N. Shore[1]

ABSTRACT
This review concentrates on observations that will be performed with the new generation of space ultraviolet spectrographs, especially the Goddard High Resolution Spectrograph on the Hubble Space Telescope. The broad astrophysical problems are outlined and the needs for specific laboratory data are placed in context. The existing laboratory data will need to be improved significantly to provide both wavelengths accurate to $\lambda/\Delta\lambda \geq 10^6$ for resonance transitions and oscillator strengths accurate to better than 10%. In addition, weak lines of ions of rare heavy elements can now be studied with S/N ratios better than 100, and the existing databases for these species will need expansion. Molecular data will also be needed for a wide range of applications.

## 2.1 Introduction

Astrophysicists act collectively like the Robin Hood of Brooks and Reiner (1981): they rob from the data rich and keep everything. The basic data obtained from laboratory investigation are essential to the understanding of the diverse cosmic environments with which observers and modelers have to deal. I speak here as a representative from the astronomical side of the business. My aim in this review is to summarize some areas of astrophysics that will be important in coming years of satellite spectroscopy and to provide pointers to where the laboratory data needs will likely be greatest.

The symbiosis between the laboratory and the cosmos that first gave birth to astrophysics is needed now more than ever. At a time when both fields are undergoing enormous changes, both in precision of measurement and breadth of spectral coverage, a closer cooperation and deeper appreciation of the respective methods and problems of the these two "cultures" is needed. But before these disciplines can fruitfully work together, they

---

[1]GHRS Science Team/CSC, Code 681, Goddard Space Flight Center, Greenbelt, MD 20771 USA

have to understand the observations and quality of the data.

The past 15 years of space astronomy have been largely defined by the broad areas of investigation pioneered in the ultraviolet by the International Ultraviolet Explorer satellite (IUE). These are certain to be the focus of the spectrographs on the Hubble Space Telescope (HST) and Lyman/Far Ultraviolet Spectrographic Explorer (FUSE/Lyman). They are: 1. planetary atmospheres, magnetospheres and comets; 2. stellar atmospheres; and 3. diffuse interstellar gas. The last of these needs to be understood in both Galactic and extragalactic contexts.

These broad categories illustrate range of problems and the diversity of the astrophysical observations. They also point out typical examples of required data, and also highlight the existing *lacunae* in the available data sets. I shall touch briefly on each of them in this review, the main purpose of which is to place the astrophysical problems in a laboratory context. To this end, I will review some of the basic problems that have been, or will be, addressed by space missions either operating, under construction, or in the final planning stages.

It should almost go without saying that the most important quantities that any laboratory study can provide are accurate energy levels and oscillator strengths or transition probabilities. These are the bread and butter of the business of remote analysis of astrophysical plasmas. The real improvement represented by the new generation of instruments, especially the Goddard High Resolution Spectrograph (GHRS), is the level of accuracy. Oscillator strengths that could be tolerated to within a factor of 2 must now be obtained to better than 10 percent. Ultraviolet wavelengths are needed to better than 0.001Å from $\lambda$1200-3200Å. The existing line lists, especially those of the heavy neutrals, and singly and doubly ionized species, for many species are only accurate to within 0.01Å at best (see *e.g.* Johansson 1988).

However, more is often required. At times, good state classifications are required in order to complete multiplets, thereby increasing the ability of the working spectroscopist to obtain complete identification of trace species. Accurate photoionization cross sections are needed for a broad range of applications, from photoionization models of H II regions to stellar interiors calculations. Collisional excitation and ionization cross sections are often needed, both for statistical equilibrium models and for detailed calculations of line strengths and for density and temperature determinations. I will try to place these laboratory data needs in perspective in the sections that follow.

## 2.2 Current and Future Space Spectroscopy Missions

At the time of the IAU meeting, there were only two working high resolution spectrographs, the International Ultraviolet Explorer satellite and the Goddard High Resolution Spectrograph (GHRS) on HST. During the near term (the next five years) there will be an additional high resolution IR instrument on the Infrared Space Observatory (ISO) and the GHRS will be replaced by the Space Telescope Imaging Spectrograph (STIS). Finally, by the end of the decade the extreme ultraviolet should be available for high resolution observation through Lyman/FUSE. This section will concentrate on the capabilities of these instruments in order to justify the laboratory data needs of space astronomers. The projects described below are common to many of the missions, and the GHRS is representative of the capabilities of the next generation as well as those of the existing instruments.

### 2.2.1 International Ultraviolet Explorer Satellite

Astronomical high resolution ($R \equiv \lambda/\Delta\lambda \geq 10000$) ultraviolet spectroscopy was pioneered by Copernicus (OAO-3) in the 1970s. Rocket-borne spectrographic experiments have achieved extremely high resolution, but have operated for only minutes at a time and on very few objects (see Jenkins *et al.* 1989). Astronomical (as opposed to planetary) far UV spectroscopic observations, below λ1000Å have been obtained for more than 15 years with the Voyager Ultraviolet Spectrometers (UVS), but these are esentially slitless and have very low resolution.

Currently the longest-lived dedicated astronomical satellite, IUE is a 45 cm telescope in geosynchronous orbit with an instrument complement of two spectrographs. One camera operates in the range 1150-2000Å, the short wavelength primary (SWP) camera, and the other operates from 1900-3400Å,the long wavelength primary (LWP) or redundant (LWR) camera. The detector is a vidicon, with 8 bit digitization. Spectra can be obtained in two modes: low resolution ($R \approx 300$) and echelle ($R \approx 10000$). There are two apertures, one 10×20 arcsec$^2$ and the other a 3 arcsec circular aperture. In low resolution mode some spatially resolved long-slit spectroscopy is possible. The limiting sensitivity is $10^{-15}$erg s$^{-1}$Å$^{-1}$cm$^{-2}$ in an -hour observing session. So far this spectrograph has produced over 75000 observations of just about every type of astronomically interesting source. Nearly all of our understanding of ultraviolet spectra of cosmic objects comes from IUE and it has certainly driven much of the need for laboratory work.

### 2.2.2 The Goddard High Resolution Spectrograph

The GHRS on the Hubble Space Telescope is the highest resolution UV spectrograph currently available for regular guest observer use. HST is a low earth orbit satellite, with a period of about 94 min. This orbit places some severe operational constraints on the instrument, some of which we will discuss below. The GHRS observes in both first order grating and echelle modes, with resolutions ranging from 800 to 90000, depending on choice of aperture and grating. The effective spectral range is 1050 to 3200Å, with the shortest wavelength being available only for the lowest order (R=2000) grating. The spherical aberration of HST degrades the image for the large science aperture (2 arcsec) resulting in a wavelength resolution lower by about a factor of 3 compared to the small aperture (0.25 arcsec). The detectors are digicon self-scanned silicon arrays. The wavelength range is limited by the photocathode sensitivity. A special advantage of the digicon is that anticoincidence circuitry permits the flagging of cosmic ray events, thereby increasing the S/N ratio of spectra taken during relatively high radiation intervals. The shortest integration time is about 100 $\mu$sec, and in rapid readout mode very short integrations can permit the study of very rapidly changing plasma conditions.

The digicon detector is also employed on the Faint Object Spectrograph (FOS) on HST, so it is worthwhile mentioning how this instrument works and some of the special features of digicon operation that allow for extremely precise measurements of wavelength and intensity. A photon liberates an electron from the photocathode, which is then accelerated through a 22 kV potential to strike the diode array. A single pulse (one count) consists of about 5700 electrons. Any cosmic ray striking the array produces only a single count for each diode it hits; the anticoincidence trigger flags Cerenkov-induced multi-diode events and also multiple cosmic ray hits. One special feature of the GHRS is its ability to remove the fixed pattern noise associated with the diode array irregularities. The diode can be oversampled up to a factor of two by electronically steering the image over the array. The same ability is also used by the flight software to correct, *in situ*, for the orbital motion of the spacecraft relative of the source. This feature, called Doppler compensation, allows the GHRS to continually shift the image during an integration to insure that any wavelength corresponds to the same diode throughout the duration of an accumulated exposure. Wavelength calibrations are provided by on-board Pt lamps, calibrated by NIST (Reader *et al.* 1990).

### 2.2.3 Space Telescope Imaging Spectrograph

STIS, currently in final planning stages, is the second generation orbital replacement instrument for the GHRS on the Hubble Space Telescope. It will have imaging detectors, so is will be able to treat a complete echelle spectrum at one exposure, in contrast to the GHRS which requires many individual order observations to cover the full range of the echelle. The spectrograph is currently designed with two types of detectors: multi-anode microchannel array (MAMA) and charge-coupled devices (CCDs), each optimized to one of two broad wavelength regions. The CCDs will cover the terrestrial UV and optical portions of the spectrum, with resolutions of approximately 800 to $10^4$ for $\lambda\lambda$3050-10000Å; this wavelength range uses only first order gratings. The vacuum UV from $\lambda\lambda$1150-3100Å is covered by the MAMA detectors. These will have resolutions of approximately 1000 and 20000 for the first order gratings, and 20000 and $10^5$ for the echelle mode. Imaging spectroscopy, a new feature of this instrument, will be an important addition to the capabilities of the instrument as will be discussed shortly.

Expected to be completed in 1997, STIS should dramatically increase the wavelength range as well as the absolute wavelength accuracy of the high resolution observations possible on HST. The single advance represented by this instrument, aside from the broad wavelength coverage at high resolution, is STIS's multiplexing capability. The GHRS is only able to observe about 8Å of spectrum at a time in the echelle mode. As with IUE, STIS is designed to image the entire echelle spectrum on a large format ($1048\times1048$ pix$^2$) array.

### 2.2.4 Lyman/Far Ultraviolet Spectrograph Explorer

Lyman/FUSE is planned for launch just after the turn of the millennium. The spectrograph is a single Rowland circle instrument with an effective area of 1200 cm$^2$ at 1100Å. The telescope is a Wolter II grazing incidence design that should provide images with full width at half maximum (FWHM) of one arcsec. Lyman is designed to provide far-UV (FUV) high resolution spectroscopy from 912 - 1500Å. There will be some overlap with the IUE and GHRS spectral range, but the primary concern is the FUV; there will also be an extreme-UV (EUV) sensitivity below $\lambda$900Å. MAMA detectors are used to cover the full range, permitting photon counting down to 100Å. The resolutions are expected to be 30000 for $\lambda\lambda$910-1250Å, 15000 for $\lambda\lambda$910-1150Å, 4000 for $\lambda\lambda$400-1600Å, and between 130 and 900 for $\lambda\lambda$100-350Å. Present plans call for a long period (24 hour) orbit with large eccentricity.

### 2.2.5 Orbiting Retrievable Far and Extreme Ultraviolet Spectrometer

ORFEUS is a spectroscopy component of the ASTRO-SPAS package which will be launched and retrieved by the Space Shuttle. Each mission is expected to last from 7 to 10 days (limited by the duration of a shuttle flight). The primary grating is a Rowland circle. The echelle mode covers from 900-1250Å with a resolution of $10^4$. The effective area is about 10 $cm^2$, and MAMA detectors are used. There is a second, lower resolution (R=5000) spectrometer for the wavelength range 390 to 1200Å.

### 2.2.6 Additional Remarks

There are a number of other missions planned for the next decade. These are briefly outlined in Kondo (1990). One infrared mission should be mentioned, however, the Infrared Space Observatory (ISO). Although primarily an infrared imager, polarimeter, and photometer, ISO also contains two spectrographs. The short wavelength spectrometer (SWS) covers the range 3-45$\mu m$ with resolutions of 1000 over the entire range and 30000 between 15 and 30$\mu m$. The low resolution mode is provided by gratings, the highest resolution is achieved using Fabry-Perot interferometers. The long wavelength spectrometer (LWS) covers the range 45-180$\mu m$ with resolutions of 200 and 10000 over the entire range. ISO is expected to be launched in 1993 into a long period, high eccentricity orbit with a minimum mission lifetime of about 18 months (limited by the supply of cryogen).

The main reason for pointing to the properties of these various instruments is to call attention to the wonderful new state of astronomical observations and modeling methods. Our ability to make precise wavelength and line profile measurements at high resolution is beginning to exceed, or has actually surpassed, the available laboratory data. With resolutions a high as $10^5$ and the possibility of achieving S/N ratios of greater than 100, species only poorly studied in the laboratory take on a whole new importance. In the rest of this review, I will outline not so much where the specific needs lie (that will be found in other reviews in this volume), but rather some of the broader problems to be addressed using this new generation of instruments. I will point out how some of the laboratory studies will have to be expanded in order to address new questions arising from these investigations.

## 2.3 Planetary Sciences: Atmospheres, Magnetospheres, Comets

For planetary atmospheres the primary need is for vibrational and electronic transition probabilities and oscillator strengths of molecular species ranging from diatoms to very complex species. In addition, van der Waals broadening coefficients are needed in order to properly calculate the atmospheric structures. Here I will mention some of the specific projects that are earmarked for GHRS, STIS, and Lyman investigations. These are primarily concerned with the energetics and composition of planetary magnetospheres and auroral zones, and with cometary physics.

Earth orbital reconnaissance over timescales of decades, a product of IUE and a promise of both HST and Lyman, will be essential in unravelling the long-term behavior of global atmospheric systems in the Jovian planets. Recent work on $H_3^+$ in the Jovian atmosphere (Drossart *et al.* 1989) has provided an important tool for the imaging and spectroscopy of auroral regions of planetary atmospheres. Accurate identification of energy levels and oscillator strengths should allow for the determination of the spectrum of precipitating magnetospheric particles.

Ultraviolet observations of cometary comae and of planetary magnetospheres have emphasized the need for additional data on $S_2$ and related sulfurous ionic species (*cf.* Festou and Feldman 1989), as well as the need for accurate transition probabilities and wavelengths for OH and even neutral atomic lines. High resolution ($R>20000$) observations will permit the analysis of both dynamics and thermal conditions in the comae and also of plasma properties of planetary magnetospheres, especially the Io torus.

Fluorescence is a complex phenomenon in comets, one that yields a great deal of information about the plasma conditions within the coma and nucleus. OH is selectively fluoresced by the solar spectrum, mainly through windows in the UV continuum. As the comet's velocity vector changes relative to the sun, different transitions are systematically excited. The detailed study of this phenomenon requires very accurate transition probabilities and wavelengths in order to determine line widths and strengths.

In the infrared, arguably the most important needs for laboratory data are in the spectra of polycyclic aromatic hydrocarbons (PAHs). These species have been identified by *in situ* mass spectroscopy from cometary fly-by missions. One important point is that there are no thorough studies of the UV-IR connections between these species, and this is something that doesn't even require the high resolution capabilities of the GHRS. Broad absorption bands of complex molecules may be important for the study of the interstellar medium and their detection in comets may have a major impact on our understanding of the complex organic molecular synthesis presumed to be occurring in cometary nuclei.

In Jovian planet auroral physics, an interesting question is the selective excitation of $H_2$ Werner band transitions shortward of 2000 Å. The relative strengths of specific electronic and vibrational transitions can be altered depending on the energy spectrum of precipitating electrons and protons from the trapped particle belts (Gladstone and Skinner 1989, Trafton *et al.* 1989). Observations of the northern auroral zone of Jupiter by the GHRS made during the Ulysses encounter with Jupiter in 1992 show strong transitions of $H_2$ between $\lambda$1500 and $\lambda$1650Å(Trafton 1992, *private communication*). These spectra and others that will certainly be taken along with the Galileo encounter provide global (HST) and local (near Jovian encounters)

## 2.4 Stellar Astrophysics

The bulk of stellar problems come from the study of atmospheric abundances and dynamics. The photospheric and chromospheric spectra are our only windows into the interior of the star, and are also the fossil record of the state of the galaxy at the time of the star's formation. There are a number of problems that have been outstanding in stellar astrophysics for most of the history of space observation. With IUE, it is possible to obtain a broad, relatively low resolution, view of the UV spectrum. In many ways, these observations profoundly altered our view of stars, especially concerning their atmospheric energetics and dynamics. However, most of the study of stellar abundances remained the province of optical observers, who were routinely able to get higher resolution observations than could be hoped for on faint stars with IUE.

This situation has changed with the launch of HST and the GHRS.

### 2.4.1 Stellar Abundances and Nucleosynthesis

Nuclear processing in novae takes place on the surface of a white dwarf whose mass determines the chemical composition of the surface layers. In the case of the most massive stars, near the Chandrasekhar limit, oxygen burning produces large overabundances of Ne. During the nebular stage of the ejecta expansion, when the matter becomes optically thin and highly ionized, it is possible to use photoionization codes to compute the abundances of the nuclear products. At present, such studies are hampered by incomplete laboratory data for the spectra of Mg and Ne for higher ions than doubly ionized (typically, Ne V dominates the early optically thin stage of the expansion, but higher ions have been detected in some novae). Term assignments and collision strengths as well as recombination coefficients are still not well known for these and other light ions. Atomic parameters for coronal lines of many species are also not well determined. One reason why these species are so important is that the envelopes of many evolved objects are very inhomogeneous so that the determination of abundances rests squarely on the observer's ability to sort out the differences between excitation and abundance. Non-LTE effects dominate the line formation in these rarified environments. Novae and planetary nebulae are illuminated by very strong central UV sources. Most of their lines are formed by recombination and often come from highly excited states. These are not well studied in the laboratory. Specifically, the data for Ne IV-VII and Mg IV-VII need to be expanded.

Many of the most important lines of rare (that is, trace) species are found in this wavelength region, but are observable only with high enough S/N ratio and high wavelength resolution. The abundances are, of course, best determined using resonance transitions and these are usually available only in the UV. Most of the stars for which there is sufficient continuum

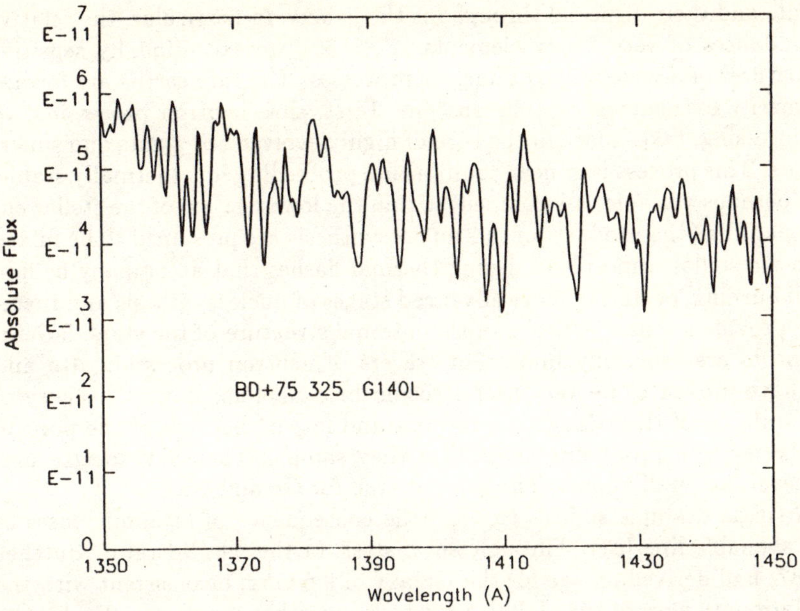


FIGURE 2.1. GHRS low resolution spectrum of the hot subdwarf BD+75°325. Notice the numerous strong lines formed in this hot, high surface gravity, stellar atmosphere.

intensity and yet not too high an ionization are between 8000 and 15000 K, mainly A and B stars. For these stars, the ions are primarily singly and doubly ionized.

Recently it has become possible to study the abundance of heavy metals in extremely evolved, post-red giant, stars, the hot subdwarfs. These stars often have temperatures greater than 30000 K (like BD+75°325 and G191B2B) (see Fig. 1). The main ionization stages of the heavy elements are between four and six times ionized. The spectra of Fe V-VII have not been well characterized in the UV, even to the extent that multiplets are not available for most of these ions and it is not possible to model their contributions to the UV spectrum. Some recent work has been able to fill in at least line identifications, but the additional quantitative work required to understand the abundances using these ions is still lacking. The iron peak is especially important here because the metal abundance is presumed to come, in part, from the accretion of material from the interstellar medium (ISM). Previously thought to be metal deficient because of the lack of lines, the spectra of these very hot objects actually consist of many weak lines of these high ions.

Several outstanding questions in this field concern the origin of the elements and their dispersal throughout the galaxy. In particular, the relative abundances of very heavy elements ($Z > 30$) are exceptionally sensitive signatures of a wide range of nuclear processes. The rare earths are formed primarily by neutron capture on $^{56}$Fe. First, slow neutron processing, or s-processing, takes place in the core of highly evolved red giants and supergiants. This process is in near equilibrium, proceeding in a thermally stable, but perhaps variable, environment within the lowest layers of the stellar envelope. Then the products of this nucleosynthesis are presumed to be mixed into the stellar atmosphere during thermal flashes that accompany helium shell burning, or during more advanced stages of nucleosynthesis and therefore provide a sensitive probe of the internal structure of the stars. Several elements are especially important tracers of neutron processing. Ru and Os have proven to be very useful cosmochronometers, in part because of the stability of their daughter products and in part because of the portion of the neutron processing chain that they sample. Their UV spectra are, however, not well known. The same is true for Ge and Ga.

A single example suffices to show the consequence of incompleteness of the available line lists. Through an analysis of the Th/Nd ratio, Butcher (1987) had derived an age for the Galaxy of 9.6 GYr, inconsistent with the evolutionary ages of the globular clusters and other population II objects. This study has sparked several enthusiastic laboratory practitioners to reexamine the spectroscopic data usually used by astrophysicists (Lawler, *et al.* 1990, Learner and Thorne 1991). The most important transition for Th II is $\lambda$4019.129Å. Lawler *et al.* (1990) showed, in a reanalysis, that Co I $\lambda$4019.126Å makes a major contribution to the Th II line strength, thereby reducing the Th abundance and increasing the age of the Galaxy to 15-20 GYr. A combination of laboratory (laser-induced fluorescence) and astronomical (solar spectral) analyses led to a determination of the Co I oscillator strengths and the final abundance for both Co and Th. Similar work will be required for other important species, especially Os and Ru. Learner and Thorne (1991) showed how modern FTS methods substantially improve the accuracy of wavelengths for even very weak transitions. In particular, they examined the wavelengths for the known constituents of the $\lambda$4019Å blend as a way of highlighting the complex nature of many features in cool stellar spectra and the need for improvement of the database accessible to astronomers[2].

There are several other ways of producing neutron rich species, mainly of light elements. In the normal carbon cycle, $^{12}$C is converted to $^{13}$C by a combination of proton capture on the parent nucleus and subsequent $\beta$ decay of the $^{13}$N nucleus. If the site is not thermally stable, and is also con-

---

[2]In particular, Learner and Thorne state that "...such a database should be free from the problems caused by unsuitable sources, inadequate standards, eye estimates of intensity, non-existent levels, refurbished measurements that prove to be 50 or 100 years old ...".

vective, it is possible to mix some products of this disequilibrated reaction up into the atmosphere. This is observed directly in evolved giants, where the $^{13}C/^{12}C$ ratio can be as large as 1/3 (the terrestrial value is about 1/90). A very important laboratory measurement is the direct determination of ultraviolet wavelengths and oscillator strengths for simple molecules like CH, CN, and CO for the various isotopic compositions. The bulk of the measurements are now performed in the optical and near IR, but in the UV it is possible to observe the ground state of the various transitions (for instance, the CO overtone at about $2\mu m$ is well studied in many stars but it is subject to problems connected with the line formation that the $A^1\Pi$-$X^1\Sigma^+$ series of CO is not).

The only information we have about the chemical evolution of the Galaxy over the full span of its history comes from the analysis of stellar atmospheric spectra. Here the need is most acute for the heavy elements. Neutral and ionized spectra for the iron peak and beyond are still very incomplete, leading both to numerous unidentified lines and to possible confusion in the identification and abundance determination of trace species of nucleosynthetic importance (see Leckrone *et al.* (1991) for a striking example in the spectrum of the chemically peculiar star $\chi$ Lupi) (see Fig. 2). The site of r-process nucleosynthesis is still an unsettled matter, and the sorting out of s- *vs.* r-process elemental abundances in stars, from Ba and S stars through the chemically peculiar main sequence stars, is an important clue. Therefore, the rare earths are very important, although the lack of detailed laboratory analysis of many of these species still hampers serious progress in this field. In most of the stars that display anomalously strong lines of these elements, the atoms are primarily doubly ionized. This means that good UV laboratory studies are urgently required in order to determine abundances. Light ions are most important for understanding nucleosynthesis both cosmologically and in intermediate and low mass stars in very advanced evolutionary stages. Available instruments can now routinely obtain R>80000 in the $\lambda\lambda$1200-3000Å region, seriously pushing the incompleteness of available laboratory identifications.

For statistical purposes, currently avalable line lists are quite good. For example, for calculating mean opacities as a function of temperature and density the large compilations and model calculations are just fine (see, for instance, the discussion by Kurucz (1992). There are many applications, however, where the details are very important and uncertainties in the available atomic parameters just don't suffice. For instance, take the case of a circumstellar shell. Many stars undergo episodic mass loss. The physical mechanism that causes a specific type of variation is often not well understoood. I mention, just as a simple example, the outbursts of massive highly evolved supergiant stars called Luminous Blue Variables (LBVs). These stars undergo outbursts that cause their optical brightness to increase by upwards of a hundredfold and for strong blueward displaced absorption components to be displaced toward shorter wavelengths. The

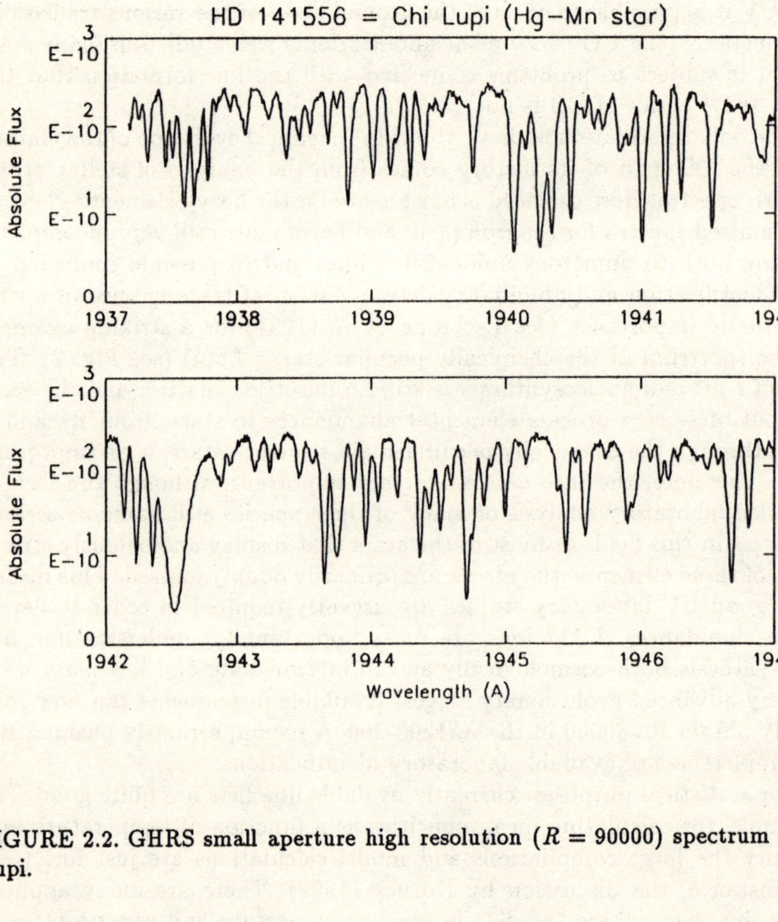


**FIGURE 2.2. GHRS small aperture high resolution ($R = 90000$) spectrum of $\chi$ Lupi.**

cause of the photometric change is that the absorption by neutral and singly ionized iron peak elements, the so-called "iron curtain", completely blankets the ultraviolet and redistributes the incident surface flux, at the base of the ejected shell, into the visible. The stars are intrinsically quite hot, often with surface temperatures $\geq$ 20000K. Now in order to determine the mass loss rate, one needs to be able to measure the turbulent broadening, excitation teemperature, and column density for the ejecta. Here details in the line lists are important because the lines are so saturated, due to mass loss rates as high as $10^{-3}M_{\odot}yr^{-1}$, where normally the mass loss would be about two orders of magnitude lower than this. The instrinsically weak atomic lines are the only unsaturated features in the spectrum. It is usually these lines, however, coming from excited states that are often hard to study in the laboratory, that have wavelengths and oscillator strengths that are the most poorly known. It is not that any single transition is especially important, unlike the situation in the GHRS spectrum of $\chi$ Lupi. Rather, any port in a storm will do when trying to obtain physical estimates for dynamical and abundance properties in these stars.

Detailed non-LTE analyses of stellar chromospheres and atmospheres require model atoms of increasing complexity for statistical equilibrium calculations. Collision strengths, transition probabilities, state term assignments, and energies are needed for neutrals and singly and doubly ionized elements ranging from C, N, and O through the heavy metals such as Hg and Pt. Isotopic shifts and hyperfine structure analyses are needed for several heavy species, notably the rare earths. Some of these data have recently become available through the analysis of Pt (Reader *et al.* 1990) but other analyses are required. The Opacity Project (Mendoza, *these proceedings*) has gone a long way toward making available collision strengths for light ions, but laboratory work on heavy species is also needed. Also needed are excited state identifications for molecules, notably oxides and nitrogen-bearing molecules. Identification of lines of highly ionized metals, like Fe V - VII are important for hot subdwarf and Wolf-Rayet stellar atmospheres (Feibelman and Bruhweiler 1990, Hubeny, Altner, and Heap 1991). As an example, in several Wolf-Rayet star spectra, especially the binary system CQ Cep, there is a strong line at $\lambda$1505Å that remains unidentified. It may simply be a blend of lines from Fe V or such like ions, or it may be due to an excited state of a doubly or triply ionized species with an excitation of about 30-40 eV but is otherwise unknown[3].

Oscillator strengths and multiplet assignments for lines of Fe II still contribute the dominant uncertainties in modeling novae and other massive ejecta from stars, like luminous blue variables and supernovae. Ionized iron peak line opacity appears to be important in structuring the emergent spec-

---

[3]Stickland *et al.* (1984) have attempted to assign an excitation to this line on the basis of its velocity in the wind of the WR star in the CQ Cep system. Several other systems show this line as well.

trum from a wide variety of stars, representing one of the primary absorbers in the surrounding smog. In fact, mainly because of the thermal regulation provided by the hydrogen Balmer lines, especially H$\alpha$, the typical temperature of astrophysical plasmas is about 10000K so the singly ionized iron peak elements (20$\leq$ $Z$ $\leq$28) are ubiquitously present in emission and/or absorption in the spectra of cosmic objects. For instance, the spectrum of $\alpha$ Ori shows not only CO absorption from the circumstellar envelope in the UV spectrum but also numerous blends of lines of ionized iron peak species. The same is true for symbiotics (Shore and Aufdenberg 1991, *in preparation*). Without complete laboratory data for these transitions, it is not possible to determine the extent to which the underlying stellar spectrum has been altered by the environment; the combined effects of circumstellar molecular *and* atomic absorption can be even more profound, as in $\alpha$ Ori (see Fig. 3).

Another illustration of how a single multiplet of a dominant atomic species can be important is provided by the intermediate resolution (R = 15000) observations of the K giant $\alpha$ Tau (Carpenter *et al.* 1991). One 35 Å region was obtained of the C II $\lambda$2320 Å multiplet, four of the lines having been previously identified with IUE. In order to detail the precise structure of the chromosphere of this M supergiant, it is necessary to have a complete set of the lines of the multiplet with which to form flux ratios. The remaining unobserved line was easily detected with the GHRS. This multiplet is an important thermostat of the lower chromosphere of cool stars and is much stronger in the giant than in the Sun. With the completion of the multiplet, more detailed modeling of the chromosphere is possible covering a wider range of column densities and temperature. However, the collision strengths and oscillator strengths for this multiplet have never been measured in the laboratory.

The analysis of silicon (Artru *et al.* 1981, Artru *et al.* 1989) has been an important tool in the modeling of chemically peculiar stars and also provides a sterling example of how such work, done with an eye to the needs of the observers, can dramatically assist the understanding of complex problems.

### 2.4.2 Mass Outflows

The problem of the line driving of stellar winds is one that illustrates many of the astrophysical problems associated with opacity. Radiation pressure is thought to be responsible for driving the winds of massive stars. The mechanical luminosity, $L_W = \frac{1}{2}\dot{M}v_\infty^2$, is usually a relatively small fraction (of the order of a few percent) of the total radiative luminosity, $L$, of the star. Here $\dot{M}$ is the mass loss rate and $v_\infty$ is the wind terminal velocity. Radiative driving occurs because of the fine tuning of the photospheric and sub-photospheric photon spectrum with respect to the distribution of lines from the atmospheric constituents. Specifically, the radiation pres-

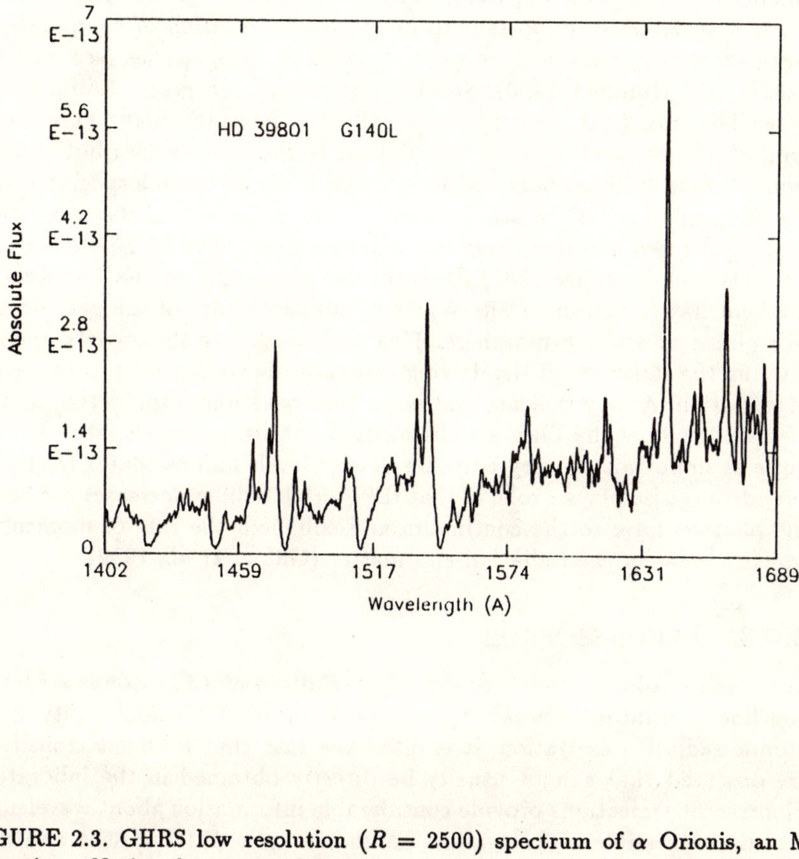


**FIGURE 2.3. GHRS low resolution ($R = 2500$) spectrum of $\alpha$ Orionis, an M supergiant. Notice the strong absorption system of CO (A-X) and complex absorption and emission structure of the continuum.**

sure produces an acceleration, given by $g_{rad} \sim \int \kappa_\nu F_\nu d\nu$, where $F_\nu$ is the monochromatic flux and $\kappa_\nu$ is the ionization- and abundance-dependent opacity. This last definition means that the opacity is summed over all contributors. A spectral distribution that maximizes the flux at the peak of the opacity near the stellar surface produces the strongest mass outflow.

The relevance of laboratory astrophysics to this problem comes from the need to determine $\kappa_\nu$. Here, as in all other atmospheric transfer problems, it is centrally important to know the distribution of comparatively weak atomic transitions from excited states of trace species (see also Kudritzki and Hummer 1990). Specific ions are of the greatest importance here. The iron peak elements, especially the neutral, singly, and doubly ionized species, are the prime contributors to the opacity distribution function (Kurucz 1991, Abbott and Lucy 1985). This enters in a specific way in the determination of the atmospheric albedo. A portion of the total radiation is scattered and this alters the underlying structure of the photosphere (Abbott and Hummer 1985). In turn, the steepening of the temperature gradient has the effect of altering the ionization state of the gas, relative to a quiescent static atmosphere. This feeds back into the opacity, thereby altering the structure of the driving region of the stellar wind at the depth of initiation. A very delicate and important question is the determination of precisely how the lines are distributed. Accurate wavelengths are not quite as important as very accurate energy levels and oscillator strengths. Broadening also plays a role, in that the overlap of lines decreases the access the photons have to the continuum and enhances the rate of momentum transfer between the radiation and matter (Gabler *et al.*, 1989).

### 2.4.3 FLUORESCENCE

Very high resolution spectroscopy is especially useful for atomic structure and line identification in late-type stars. Because of the nonlocality of the atomic radiative excitation, it is often the case that resonant transitions are observed that cannot usually be directly obtained in the laboratory. Fluorescent transitions provide considerable information about wavelength regions inaccessible to observation. For instance, the O III $\lambda$3130Å multiplet is pumped by He II Ly$\beta$, the Bowen fluorescence mechanism. Observation of these transitions aids the determination of the far UV continuum of hot stars, especially in H II regions.

Willson (1975) first suggested the role of Fe I fluorescence in the formation of late-type stellar spectra, especially T Tauri stars. Since then, many species, especially Fe II, have been identified from their fluorescence spectra and have become important tools for understanding the thermodynamics of cool stellar atmospheres (Carpenter and Johansson 1988, Johansson and Carpenter 1988, Carpenter *et al.* 1989). The ultraviolet multiplets couple to produce optical emission lines, and also can be used to identify high lying states that normally collisionally de-excite under laboratory conditions.

Fe II is also an important constituent in the spectra of active galaxies, in particular Seyfert galaxies and quasars (*e.g.* Wills *et al.* 1985), and it is responsible for much of the continuum structure observed in these objects. Far from being just a spectroscopic curiosity, fluorescent transitions provide important diagnostics for stellar atmospheres, in particular placing limits on the velocity gradient in the deep chromosphere (since decoupling takes place if the transitions are too severely Doppler-shifted with respect to each other); they are also important components in the total radiative losses for these atmospheres.

The O I resonance multiplet provides an illustrative example of fluorescence. The resonance transition, $\lambda$1302Å, pumps the $\lambda$1641Å line (the lower state of which is the upper state of the auroral line $\lambda$6300Å). Originally unidentified in a number of different objects, especially symbiotics and novae, this transition is a very important way of identifying the extent of the neutral component of emitting gas in late type stellar atmospheres. A number of other neutral species also show fluorescence. There are, however, several transitions of N I identified from IUE that do not appear to be pumped yet appear abnormally strong in highly ionized envelopes like RR Tel.

One of the best examples of the application of fluorescence to late-type atmospheres is provided by the GHRS observations of $\alpha$ Tau. Numerous Fe I lines were identified in the GHRS spectrum of this star, most of which are fluorescently pumped, and Co II fluorescence was confirmed (Carpenter and Wing 1986, Carpenter *et al.* 1991). These can be used to place limits on the velocity gradients in the chromosphere, the region from which the stellar wind emerges.

Fluorescence usually takes place because of accidental coincidences between atomic levels of often different species. Proper characterization of the excitation requires comprehensive lists of possible pump transitions. The main problem in current NLTE codes is to construct accurate model atoms with which to perform the radiative transfer calculations. This incompleteness is seen most starkly when looking at fluorescence. Laboratory studies of collision strengths for UV transitions are vital in this regard, as are transition probabilities with which one can calculate the densities. Many of the lines so far detected in stellar spectra are pumped by H I and He II Lyman series. Therefore, measurements of these transitions yield information about the line strengths in the far ultraviolet (FUV) and and extreme ultraviolet and soft x-rays (EUV). In addition, many emission lines that remain unidentified in chromospheric spectra are likely caused by fluorescence. The work by Feibelman *et al.* (1991) is an especially interesting attempt to use laboratory Fe II studies to analyze the role of fluorescence in a wide range of objects.

### 2.4.4 Coronal Species and Very High Ionization Plasmas

Ions characteristic of gas with temperatures greater than about $10^5$K are observed in a wide variety of stars, from late-type stars with true coronae and flare activity to novae. For dMe stars, identification of lines from these ions is important for the determination of plasma conditions during flares. An example is provided by the GHRS spectrum of the star AU Mic, obtained in a post-flare state. The Fe XXI $\lambda$1354Å line, observed during solar flares, is clearly present (I should also remark that the Cl I $\lambda$1354Å line is a fluorescent line). Fe XII is also observed in the Skylab spectra of solar flares, but has not yet been detected in other stars. Accurate collision strengths and transition probabilities are especially important for these lines, which are formed typically at temperatures of order $10^7$K, in order to determine emission volumes and flare energetics.

In systems with very strong UV continua, such as the nuclei of planetary nebulae, coronal lines are often observed. These come from transitions normally detected only in the intrinsically hot (of order $10^6$-$10^7$K) plasma of the solar and stellar coronae. In extremely low density environments, however, photoionization by a very strong FUV and EUV continuum can produce lines of extremely highly ionized species. For instance, Ar V and Ne V have been detected in many planetaries. In the case of these sources, the combination of very low density and strong FUV radiation produce substantial emission in the coronal species. In addition, power law (nonthermal) continua can often overproduce coronal species in low density environments such as symbiotic stars and even AGNs, so the presence of these lines is important as a clue to the FUV continuum and optical depth of the surrounding gas. Again, collision strengths and transition probabilities are necessary to compute the statistical equilibrium for each species and obtain information about the exciting radiation source.

In novae, such lines are often the only way to obtain abundances of important species, since the strong UV radiation ionizes all of the lower ions of nucleosynthetically important species. For instance, [Mg VII] is one of the only ways of getting information about magnesium, a product of proton capture during incomplete nuclear burning on O-rich white dwarfs during the nova eruption (Starrfield and Snijders 1989). Recent novae have also shown large overabundances of Al, and coronal lines of this element will prove increasingly important.

Many lines remain unidentified in the symbiotic stars (Penston *et al.* 1983, Aufdenberg 1991, *in preparation*). These objects are ideal "laboratories" for many forbidden transitions of highly ionized species of all flavors, since their spectra are formed by the photoionization of a red giant or supergiant envelope by a stable, very hot ($T_{eff} \geq 10^5$K) UV bright source. RR Tel has been extensively used for the study of Fe II (*e.g.* Fawcett 1988) but there are many other ions still to be understood in the spectrum of this and other symbiotics, for instance V1016 Cyg (Nussbaumer *et al.* 1981). Again, the work of identifying the lines is hampered by the lack of assignments of multiplets for many of these species. Since recombination is important for these objects, many of the unidentified lines may arise from very highly excited species (Nussbaumer 1990). To make matters worse, and to underscore how difficult and bizarre the symbiotics really are, I should add that some of the emission lines may also be contaminated by environmental absorption by iron peak ions arising in the wind of the red giant companion.

## 2.5 Interstellar Medium: Diffuse Gas and H II Regions

This is a broad field, spanning the widest range of density and excitation conditions, and it is the one that most benefits from extremely precise laboratory analysis. The ISM provides examples of the most extreme disequilibrated conditions available astrophysically, in large measure because of the very broad range of timescales dominating the thermal and mechanical properties of the medium. First, in molecular clouds, the temperatures typically range from about 10-400 K and densities are in the range $n \geq 10^3 \mathrm{cm}^{-3}$. The important processes that must be considered here are connected with diatomic and more complex molecules and are best studied in the millimeter, sub-millimeter, and infrared portions of the spectrum. Second, the diffuse interstellar medium varies in temperature between a few hundred and a few thousand Kelvins. The density is typically between 0.1 and $10^2 \mathrm{cm}^{-3}$. Finally, the hot interstellar medium, with temperatures typically greater than $10^5$K and very low density, rounds out the basic phases. There are also H II regions, ionized environments around UV-bright stars like massive OB stars and the central stars of planetary nebulae, as in Fig. 4, which shows the spectrum toward a typical Galactic plane O star.

### 2.5.1 H II Regions and Emission Line Spectra

The ions that dominate the UV emission spectrum of planetary nebulae and H II regions are He I-II, C II-IV, N II-V, O II-VI, Ne II-V, Mg II-VI, Si II-IV, Ar III-V, and S II-IV. As investigations push into the FUV, higher ions like P V become more interesting. The emission lines are formed by a variety of mechanisms, including fluorescence (pumped by the H I and He II Lyman series), recombination, and collisional excitation. Several are the result of dielectronic recombination effects, which are very important in governing the thermal state of the gas (Osterbrock 1989). Many of these species are reviewed by Mendoza (these proceedings) and are the targets of the Opacity Project studies, so here the situation is very likely to improve

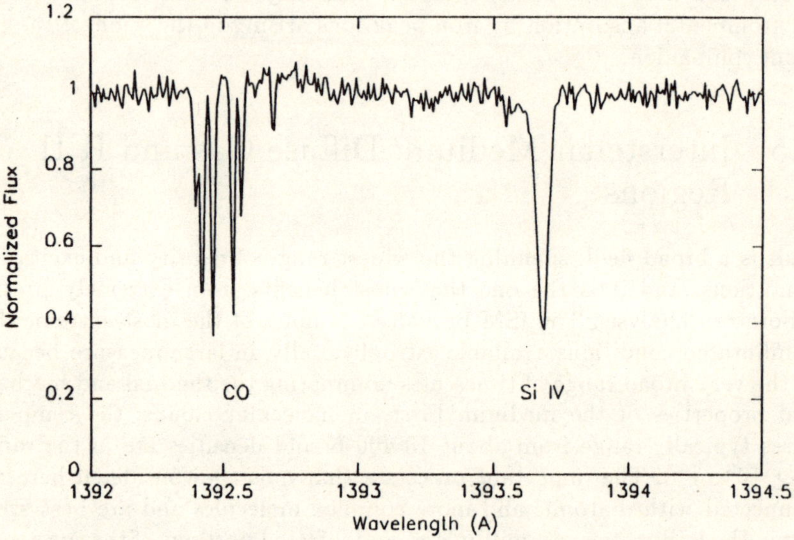


**FIGURE 2.4. Illustration of the variety of phases present in the interstellar medium. This is a GHRS echelle small aperture spectrum of ζ Ophiuchi, a galactic O star. The strong absorption lines are CO from the cool periphery of translucent clouds, and the Si IV is formed from the compact H II region surrounding the star.**

rapidly with the release of the project's database. However, benchmark laboratory measurements are still needed for these many of these species.

Photoionization codes are now available for modeling many of these transitions, mostly making use of escape probability methods to compute the statistical equilibrium and line strengths (Ferland 1990). There are many examples where the uncertainties in laboratory data limit the utility of these codes. Several important emission lines lack good transition probabilities, in particular Ne V $\lambda$1134, S III $\lambda\lambda$1713, 1728Å, and Fe III $\lambda$1122Å. The high ions of phosphorus are also not well studied. Collision strengths are especially uncertain for a number of important lines like N IV] $\lambda$1486, N II $\lambda$1084, N I $\lambda$1200, O I $\lambda$1304, [Mg VII]$\lambda\lambda$2510, 2629, S III $\lambda$1256Å, and many coronal species like Fe XII and Fe XXI (both around $\lambda$1360Å). Since photoionization is an important process for abundance determinations in a wide variety of objects, as we have already seen, these uncertainties have a ripple effect through many problems.

Factor of two errors are still present in many transition probabilities and collision rates for forbidden transitions of highly ionized species, especially for ions of the noble gases. These are among the major uncertainties in the abundances derived from emission line analyses. A case in point, although for an optical transition, is the problem of the Ni abundance in supernova remnants. Henry (1984) argued that the Ni abundance derived from an forbidden line at $\lambda$7378Å was suspect because of the uncertain transition probability for the line. Subsequent analysis by Hester *et al.* (1990) has extended this and pointed to the laboratory data, rather than the excitation conditions of the knots of emission in the remnant, as the cause of the uncertainty. Since Ni is an important supernova nucleosynthetic product, the determination of its abundance is important in understanding the details of the supernova explosion; supernova remnants are the only direct way of gaining access to these details. Many of the important lines for these heavy elements are available in the UV, but the laboratory data are still sparse (Kelly 1987).

An important feature of future spectrographs, especially STIS, will be the capacity to obtain spatially resolved spectra. Long slit spectroscopy has been important in groundbased work for some time, since it provides dynamical information over a broad region as well as information about populations (for stellar absorption line studies, for instance, of galaxies) and excitation in the case of nebulae. The multiplexing capacity of imaging detectors extends the range of questions that nebula modelers can address. For instance, the variation in the excitation conditions within an H II region is vital information for determining not only thermodynamic but also abundance variations. Spatial imaging of expanding structures allows the observer to separate individual knots of emission, for instance, in both space and velocity, providing a sort of tomographic picture of the plasma. Although we do not yet have this capability in the ultraviolet, the next generation of space spectrographs will certainly provide it. With this ability, it will be possible to model the *range* of density and pressure within H II regions, to study the turbulence, and to determine energy budgets for the gas on many diferent spatial scales. All of these questions require accurate wavelengths ($R \geq 10^5$ to match the internal velocity dispersions in the H II regions) and good transition probability values for highly excited transitions of light ions.

### 2.5.2 Absorption Line Spectra

Perhaps more than any other set of problems, spectroscopy of the interstellar medium pushes the available spectroscopic facilities to their limit. The intrinsic line widths in individual diffuse clouds are typically only a few km $s^{-1}$ so instruments must deliver resolutions $\approx 10^5$ to come close to resolving individual line profiles. The medium along any line of sight does not generally appear to consist of only one or even two clouds (Jenkins 1986, Jenkins *et al.* 1989). Given the usual distance of several kiloparsecs to the average light source, generally an O star, the line profiles consist of a very complex collection of individual contributors, each of which may have different abundances (see below), different excitation and ionization conditions, and different densities.

At the time of this writing (winter 1991), the short wavelength echelle mode ($\lambda < 1700$Å) of the GHRS is not functioning because of a failure of a low voltage regulator. It is believed that the detector is still fine. The long wavelength echelle mode is, however, still fully operational. The first order gratings, which provide resolutions of order 25000 with the small (0.25 arcsec) aperture, can be used to obtain line profiles with about a factor of 2 higher resolution than obtained using Copernicus, with greater sensitivity and no scattered light (Fig. 4). While the detailed mapping of isotopic compositions in the ISM (specifically using CO and other molecular species in the short wavelength region) will not be possible in the near future, *many* other important problems that tax the available data are still doable.

Rocket observations have been performed at extraordinary resolution, $R = 10^6$, by Jenkins *et al.* (1989) for the O star $\pi$ Sco. These spectra give a clearer view of the complexity of the ISM than any previous observations and underscore the need for applying great care in interpreting interstellar conditions and abundances.

The specific laboratory data requirements for the study of interstellar regions are quite disparate. For the diffuse interstellar medium, the primary data required are wavelengths and $gf$ values (as also discussed by Morton in these proceedings). The reason is that the collision rates are low enough that radiative excitation from the ground state followed by scattering (rather than collisional excitation/de- excitation) dominates the line formation process. Thermometry is provided by the direct measurement of fine structure levels of ground state transitions, for example the very well observed resonance transition of C II $\lambda\lambda$1334, 1335Å. Accurate oscillator and collision strengths are required to translate relative level populations into density and temperature conditions for these clouds. As stated earlier, the currently available optical and UV spectrographs are not always able to resolve the intrinsic line profile for a single cloud. Therefore, what we see in the UV is an amalgam of many individual clouds, each with its own specific internal properties. When viewing a line profile, for instance the C I** $\lambda$1193Å line in the $\xi$ Per spectrum compared with the Si II $\lambda$1193Å line, depending on the velocity of the clouds along a line of sight, we may or may not be able to separate out individual absorption regions. Making sense out of the spectrum obtained along any line of sight thus requires at least that the laboratory data exceed the ability of the instrument to resolve the line profiles.

Because the medium is very inhomogeneous and sparsely sampled along any line of sight, accurate wavelengths are required, with precisions better than the resolution of the spectrograph in question. Doublets are especially important because they provide an immediate handle on the excitation conditions in the diffuse interstellar medium. This knowledge is also important for other parts of the spectrum. For instance, the C II fine structure transition at 153$\mu m$ is also important, since it can be directly observed in the boundaries of heated molecular clouds and represents an important loss mechanism for the clouds.

For the diffuse interstellar medium, resonance and fine structure lines are of the most interest. The repeated collision of the medium with shocks, generated by stellar explosions and stellar winds, heats a substantial part of the medium to temperatures in excess of $10^6$K. In order to probe this part of the medium, accurate oscillator strengths and wavelengths are needed for the resonance transitions of very highly ionized species, like Fe X $\lambda$6375Å and [Fe XIV] $\lambda$5303Å (Hobbs 1984, Hobbs and Albert 1984). Some of these lines are accessible from the ground, using Fabry-Perot spectrometers, but the new generation of space spectrographs will make them progressively easier to map and study in detail using ultraviolet transitions.

An example of the problems posed by the current databases is the oscillator strengths of C I $\lambda\lambda$1190, 1193 and S I $\lambda$1474Å. The UV is the only portion of the spectrum with which we can obtain direct access to the low $Z$ neutral phase. Although Na I is a well studied optical line, the most important neutral atomic lines of C, N, and O are found at $\lambda < 1500$Å. With the current sensitivity of the GHRS, it is possible to study many lines of sight through the ISM using bright OB stars at a resolution of between 25000 and 90000.

Accurate wavelengths (to better than R=$10^5$) and oscillator strengths of resonance transitions for neutral through doubly ionized species are essential. Much of the periodic table remains to be explored in the ISM. While for $Z < 20$ the situation is pretty good, with the important exception of

Si I and Si II, the heavy metals are in need of more accurate wavelengths and oscillator strengths.

Recent work on the abundances of CO and its isotopic relatives has been possible using GHRS spectra (Smith *et al.* 1991, Bruhweiler *et al.*, *in preparation*).

Accurate energy levels and band strengths for CO, CN, OH, and $H_2$ are still extremely important and need improving. These probe the cold clouds and peripheral portions of molecular clouds (see Cardelli *et al.* 1991, Smith *et al.* 1991). Neutral and singly ionized atoms, not just containing C, N, and O but also heavy elements, are in need of accurate wavelengths and oscillator strengths. Here the need is broad, because these are the dominant probes of the diffuse interstellar medium. Coronal ions, especially Mg V-VII through Ne IV-V, and also intercombination and forbidden lines of the iron peak coronal species, are all needed for the analysis of H II regions.

### 2.5.3 Fluorescence in Reflection Nebulae and Herbig-Haro Objects

Fluorescence has recently been identified as a process in interstellar environments. In particular, Herbig-Haro objects, which are connected with protostars and appear to be the working surfaces of jet outflows from these stars, have been observed with IUE to show $H_2$ fluorescence (Bohm, Scott and Solf 1991). This fluorescence had been identified earlier by Witt *et al.* (1989) in reflection nebulae, both as extended red emission in the optical and as bands in the UV. The fluorescent yield is not, however, well known, and this hampers modeling efforts. The UV absorption spectrum cannot be observed directly because the line of sight column densities are often too low. Also, in reflection nebulae, the emission occurs *behind* the exciting source, so that the UV absorption spectrum must be calculated and compared with the excitation observed in the individual emission bands.

It is possible that other examples of such fluorescence exist – for instance, for more complex molecules, as was suggested long ago for the origin of the near IR emission PAH candidates. Ultraviolet absorption bands of these species are important for understanding the radiative excitation mechanisms involved with these otherwise poorly identified transitions (Leger and Puget 1989), and any laboratory work will be useful in helping to clear up the origin of these bands. Some recent progress has been made by measuring electron energy loss spectra at low and intermediate energies (5 - 100 eV) (Khakoo *et al.* 1990), but much more remains to be done.

## 2.6 Coda

There is a wide frontier stretching ahead and, as Schulberg once put it in the film "On the Waterfront", everybody works!

*Acknowledgements:* I wish to thank the organizers of this meeting, Peter Smith and Wolfgang Wiese, for their kind invitation to present this talk at the IAU Joint Commission meeting and to write a summary of it for these proceedings. I also thank Tom Ake for very helpful comments on the manuscript and Bill Martin, Wolfgang Wiese, Alfons Weber, Ara Chutjian, Joe Ajello, Bill Parkinson, Fred Bruhweiler, Sumner Starrfield, Glenn Wahlgren, Sveneric Johansson, and Dave Leckrone for discussions.

Abbott, D. C. and Hummer, D. G. 1985, *ApJ*, **294**, 286.

Abbott, D. C. and Lucy, L. B. 1985, *ApJ*, **288**, 679.

Artru, M.-C., Borsenberger, J., and Lanz, T. 1989, *A&AS*, **80**, 17.

Artru, M. C., Jamar, C., Petrini, D., and Praderie, F. 1981, *A&AS*, **4**, 171.

Bohm, K-H., Scott, D. M., and Solf, J. 1991, *ApJ*, **371**, 248.

Brooks, M. and Reiner, C. 1981, *The 2000 Year Old Man* (NY: Warner Books).

Carpenter, K. G. and Johansson, S. 1988, in *Advances in Astrophysics* (ed. E. Rolfs) (ESA SP-281), p. 349.

Carpenter, K. G., Pesce, J. E., Stencel, R. F., Brown, A. Johansson, S., and Wing, R. F. 1988, *ApJS*, **68**, 345.

Carpenter, K. G., Robinson, R. D., Wahlgren, G. M., Ake, T. B., Ebbets, D. C., Linsky, J. L., Brown, A., and Walter, F. M. 1991, *ApJL*, **377**, L45.

Cardelli, J. A., Savage, B. D., Bruhweiler, F. C., Smith, A. M., Ebbets, D. C., Sembach, K. R., and Sofia, U. J. 1991, *ApJL*, **377**, L57.

Drossart, P. *et al.* 1989, *Nature*, **340**, 539.

Fawcett, B. C. 1988, *Atom. Nucl. Data Tables*, **40**, 1.

Feibelman, W. A. and Bruhweiler, F. C. 1990, *ApJ*, **357**, 262.

Feibelman, W. A., Bruhweiler, F. C., and Johansson, S. 1991, *ApJ*, **373**, 649.

Ferland, G. 1990, *CLOUDY: OSU Special Report 89-01.*

Festou, M. and Feldman, P. 1989, in *Exploring the Universe with the IUE Satellite* (ed. Y. Kondo) (Dordrecht: Kluwer) p. 101.

Gabler, R., Gabler, A., Kudritzki, R. P., Puls, J., and Pauldrach, A. 1989, *A&A*, **226**, 162.

Gladstone, G. R. and Skinner, T. E. 1989, in *Time-Variable Phenomena in the Jovian System* (eds. Belton, M., West, R., and Rahe, J.) (NASA SP-494), p. 221.

Henry, R. B. C. 1984, *ApJ*, **281**, 644.

Hester, J. J., Graham, J. R., Beichman, C. A., and Gautier, T. N. III 1990, *ApJ*, **357**, 539.

Hobbs, L. M. 1984, *ApJ*, **280**, 132.

Hobbs, L. M. and Albert, C. A. 1984, *ApJ*, **281**, 639.

Hubeny, I., Altner, B. A., and Heap, S. R. 1991, *ApJL*, **377**, L33.

Jenkins, E. B. 1986, *ApJ*, **304**, 739.

Jenkins, E. B., Lees, J. F., van Dieshoeck, E. F., and Wilcots, E. M. 1989, *ApJ*, **343**, 785.

Johansson, S. and Carpenter, K. G. 1988, in *Advances in Ultraviolet Astronomy* (ed. E. Rolfs) (ESA SP-281), p. 361.

Johansson, S. 1988, in *IAU Colloq. 94: Physics of Formation of Fe II Lines Outside LTE* (ed. Viotti, R., Vittone, A., Friedjung, M.) (Dordrecht: Reidel).

Kelly, R. L. 1987, *J. Phys. Chem. Ref. Data*, **16** (Suppl. 1), 1.

Khakoo, M. A., Ratliff, J. M., and Trajmar, S. 1990, *J. Chem. Phys.*, **93**, 8616.

Kondo, Y. (ed.) 1990, *Observatories in Earth Orbit and Beyond* (Dordrecht: Kluwer).

Kudritzki, R. P. and Hummer, D. G. 1990, *Ann. Rev. Astr. Ap.*, **28**, 303.

Kurucz, R. L. 1991, in *Stellar Atmospheres: Beyond Classical Models* (eds. Crivalini, L. and Hubeny, I.) (Dordrecht: Kluwer) p. 441.

Lawler, J., Whaling, W., and Grevesse, N. 1990, *Nature*, **346**, 635.

Learner, R. C. M., Davies, J., and Thorne, A. P. 1991, *MNRAS*, **248**, 414.

Leckrone, D., Wahlgren, G., and Johansson, S. 1991, *ApJL*, **377**, L37.

Leger, A., and Puget, J.-L. 1989, *Ann. Rev. Astr. Ap.*, **27**, 161.

Nussbaumer, H. 1990, in *Evolution in Astrophysics* (ed. E. Rolfs) (ESA SP-310), p. 87.

Nussbaumer, H. and Schild, H. 1981, *A&A*, **101**, 118.

Osterbrock, D. 1989, *Astrophysics of Gaseous Nebulae and Active Galactic Nuclei* (Sacramento: University Books).

Penston, M. V., Benvenutti, P., Cassatella, A., Heck, A., Selvelli, P., Macchetto, F., Ponz, D., Jordan, C., Cramer, N., Ruefner, F., and Manfroid, J., 1983, *MNRAS*, **202**, 833.

Reader, J., Acquista, N., Sansonetti, C. J., and Sansonetti, J. E. 1990, *ApJS*, **72**, 831.

Stickland, D. J. *et al.*, 1984, *A&A*, **143**, 45.

Smith, A. M., Bruhweiler, F. C., Lambert, D. L., Savage, B. D., Cardelli, J. A., Ebbets, D. C., Lyu, C.-H., and Sheffer, Y. 1991, *ApJL*, **377**, L61.

Starrfield, S. and Snijders, M. A. J. 1989, in *Exploring the Universe with the IUE SAtellite* (ed. Kondo, Y.) (Dordrecht: Kluwer), p. 377.

Trafton, L., Carr, J., Lester, D., and Harvey, P. 1989, in *Time-Variable Phenomena in the Jovian System* (eds. Belton, M., West, R., and Rahe, J.) (NASA SP-494), p. 229.

Welty, D. E., Hobbs, L. M., and York, D. G. 1991, *ApJS*, **75**, 425.

Willson, L. A. 1974, *ApJ*, **191**, 143.

Wills, B. J., Netzer, H., and Wills, D. 1985, *ApJ*, **288**, 94.

Witt, A. N., Stecher, T. P., Boronson, T. A., and Bohlin, R. C. 1989, *ApJL*, **336**, L21.

# 3

# Atomic Data Needed for Far Ultraviolet Astronomy with HUT and FUSE

**Jeffrey L. Linsky**[1]

ABSTRACT I will summarize the spectroscopic capabilities of existing and planned space experiments, including HUT and FUSE, that will obtain spectra of astronomical sources at wavelengths shorter than Lyman-$\alpha$. The important atomic and molecular data needed to analyze far and extreme ultraviolet spectra that will be obtained with these instruments include accurate wavelengths, oscillator strengths, photoionization cross sections for six important molecules, and, especially, electron collisional excitation cross sections for both low and high stages of ionization.

It is a capital mistake to theorize before one has data.
*Sir Arthur Conan Doyle*

## 3.1 Introduction

Each time astrophysicists use more sensitive or higher resolution spectrometers to study astronomical sources, they obtain spectra that cannot be adequately interpreted without new atomic and molecular data. In this paper I will summarize what I believe are the important types of atomic and molecular data that astrophysicists need to analyze spectra that are becoming available at wavelengths shorter than Lyman-$\alpha$. I divide this wavelength interval into the far ultraviolet (FUV, 912–1250Å) and the extreme ultraviolet (EUV, 100–912Å). The Lyman continuum edge forms a natural division between these two spectral regimes, because near 912Å the instrumental techniques, interstellar absorption, and the plasma temperatures at which the important diagnostics are formed change their character appreciably.

Until recently our only glimpses of this spectral region were provided by either solar instruments on the Orbiting Solar Observatories, OSO-6 and

[1] Joint Institute for Laboratory Astrophysics, National Institute of Standards and Technology and University of Colorado, Boulder, Colorado 80309–0440 USA

-7, Skylab, the Solar Maximum Mission (SMM), and the Naval Research Laboratory High Resolution Telescope and Spectrograph (HRTS), or by relatively insensitive astronomical spectrometers on Copernicus and sounding rockets. This situation changed with the very sensitive low resolution spectrograph on the the Hopkins Ultraviolet Telescope (HUT), which flew on the December 1990 Astro-1 mission and is scheduled to refly in 1993. Spectrometers on sounding rockets, the Extreme Ultraviolet Explorer (EUVE) mission, and the Orbiting and Retrievable Far and Extreme Ultraviolet Explorer (ORFEUS) mission will provide important new spectra in the FUV and EUV for analysis. Later, the Far Ultraviolet Spectrograph Explorer (FUSE) will extend our spectroscopic capabilities to higher spectral resolution and to a wider range of sources. Thus it is now important to identify the most crucial atomic and molecular data that are and will be needed to understand high-resolution spectra in the FUV and EUV in the hope that experimentalists and theoreticians will attempt to fill the data gaps.

In preparing this talk I have asked a number of astronomers who regularly use atomic and molecular data in this spectral region for their advice concerning the major data gaps they have encountered. I will also draw heavily on the recommendations of the NASA Laboratory Astrophysics Workshop (J.M. Ajello, private communication) held at the Jet Propulsion Laboratory on 8–9 November 1989. What follows is of necessity an incomplete list, but I hope that it contains most of the important needs that can be identified at this time.

## 3.2 FUV and EUV Spectrometers for Astronomy in the 1990s

The International Ultraviolet Explorer (IUE) and the Hubble Space Telescope (HST) spectrographs, the Faint Object Spectrograph (FOS) and the Goddard High Resolution Spectrograph (GHRS), retain some sensitivity down to $\lambda \approx 1150$Å. The low reflectivity of all known materials, however, severely restricts spectroscopic observations of astronomical sources at shorter wavelengths, unless the instruments contain very few reflections from specially coated surfaces or use glancing incidence reflection, or both techniques together.

Until recently, our only glimpses of spectra of astronomical sources other than the Sun came from the Copernicus satellite, the Voyager Ultraviolet Spectrometers (UVS), and a few sounding rocket experiments. The Copernicus satellite was able to observe brighter sources in the FUV by using a 90-cm normal incidence telescope and a prime focus spectrograph with all surfaces coated with LiF. It could not track sources fainter than about magnitude 6 or 7. Its characteristics, and those of the other instruments that I will describe, are summarized in Table 1. Despite its low sensitivity and single channel photomultiplier detector system, Copernicus provided a wealth of new information on the interstellar medium through its high-resolution studies of the electronic bands of $H_2$ and HD, the O VI resonance lines, and the many other ions that show rich FUV spectra. Among its major accomplishments (cf. review by Spitzer and Jenkins 1975) were the measurement of the D/H ratio from the higher Lyman lines, metal depletion in the interstellar medium (ISM), temperatures in interstellar clouds, and the superionized plasma in hot star winds.

The Ultraviolet Spectrometers on the Voyager 1 and 2 spacecraft have observed more than 300 objects including young O and B-type stars, hot white dwarfs and other degenerate stars, active binaries, galaxies, and diffuse emission sources such as supernova remnants. The spectrometers have Wadsworth mount objective gratings overcoated with platinum to observe the 525–1650Å spectrum with $\sim$ 18Å resolution in first order, but with rather low sensitivity. For a description of the instrument see Broadfoot *et al.* (1977), and for a summary of the EUV observations obtained with these spectrometers see Holberg (1991). In many cases these low resolution spectra have been most useful in discovering which targets have significant FUV and EUV emission, so that they may be studied later at high resolution with newer instruments.

TABLE 1. Spectroscopic Capabilities in the FUV and EUV

| Spacecraft or Instrument | Dates Operational | Spectral Range (Å) | Spectral Resolution ($\lambda/\Delta\lambda$) | Effective Area ($cm^2$) |
|---|---|---|---|---|
| Copernicus | 1972–80 | 912–3000 | 20,000 | – |
| Voyager UVS | 1977– | 525–1650 | $\sim$ 70 | $\sim$ 0.1 |
| IMAPS | 1984– | 1003–1175 | 130,000 | 3.8 |
| HUT | 1990– | 425–1850 | 300-600 | 6.5 |
| Colorado Rocket | 1988– | 910–1170 | 500 | $\sim$ 1 |
| EUVE | 1992– | 80–800 | 250-500 | $\sim$ 0.5 |
| ORFEUS | 1993– | 390–1200 | $\sim$ 7,000 | $\sim$ 6 |
| FUSE | 2000?– | 910–1260 | 30,000 | $\sim$ 50 |
| | | 400–1600 | 500 | 40 |
| | | 100–350 | $\sim$ 2,200 | $\sim$ 5 |
| | | 100–350 | $\sim$ 800 | $\sim$ 10 |

### 3.2.1 Instruments for the Early 1990s

During the December 1990 flight of the Shuttle Astro-1 mission, the Hopkins Ultraviolet Telescope (HUT) obtained high signal/noise spectra of many types of sources including the Io torus (Davidsen *et al.* 1991), Comet Levy (Feldman *et al.* 1991), several types of stars, supernova remnants (Blair *et al.* 1991), stellar binaries with mass exchange, starburst galaxies, and quasars. A reflight in 1993 is planned. HUT contains a 90-cm f/2 telescope with an iridium coated primary mirror for significant reflectivity at wavelengths as short as 425Å. The Prime Focus Spectrograph has only one reflection (a grating) to minimize reflectivity losses. The spectral resolution is ~3.0Å in first order (850–1850Å) and ~1.5Å in second order (425–925Å). For a more complete description of HUT and the Astro-1 mission see Davidsen *et al.* (1991) and Blair and Gull (1990). HUT is extending our understanding by providing FUV and EUV spectra of very faint sources for the first time.

The Extreme Ultraviolet Explorer (EUVE), scheduled for an early 1992 launch, will have grazing incidence optics and variable line-space grating spectrometers to study the 80–800Å spectral band with moderate resolution (250–500). The long exposures available with this instrument will provide high signal/noise spectra of perhaps several hundred sources during the lifetime of EUVE. Thus the EUVE spectrometer will be able to study many more sources and at shorter wavelengths than HUT. It is expected that the EUVE spectrometers will observe many active stars such as M dwarfs, hot white dwarfs, and binary systems such as the RS CVn systems and cataclysmic variables. A more complete description of the instrument and simulated spectra can be found in Bowyer and Malina (1991).

ORFEUS will provide moderate resolution spectra, $\lambda/\Delta\lambda \sim 7000$, for point sources between 390 and 1200Å. This instrument will fly on the ASTRO-SPAS platform to be launched from Shuttle in February 1993 for a 7–10 day mission. ORFEUS will provide the first moderate resolution spectra of sources other than the Sun in the EUV spectral range. ORFEUS consists of four spherical diffraction gratings with variable line spaces in a modified Rowland mount. The gratings and the 1-m normal incidence telescope are coated with iridium. This instrument is described by Hurwitz and Bowyer (1991).

The highest spectral resolution obtained so far in the FUV has been with the Interstellar Medium Absorption Profile Spectrograph (IMAPS) instrument which has flown several times on sounding rockets and is also scheduled to fly on the February 1993 ASTRO-SPAS mission. This instrument, which is described by Jenkins *et al.* (1989), has achieved a resolution of $\lambda/\Delta\lambda \geq 1.3 \times 10^5$ during a sounding rocket observation of the hot star $\pi$ Scorpii. In order to utilize the IMAPS fully in measuring the velocities of interstellar clouds, an instrument with this spectral resolution requires laboratory wavelengths of atomic and molecular lines (primarily the Lyman and Werner bands of $H_2$ and HD) that are more accurate than 1/10 the instrumental spectral resolution .

Cash *et al.* (1989) have developed a very efficient conical diffraction grating spectrometer instrument that has now been flown on several sounding rockets (e.g. Green, Cash, and Stern 1991). This instrument presently provides moderate spectral resolution in the 910–1170Å band.

### 3.2.2 The Far Ultraviolet Spectrograph Explorer

After its launch, now anticipated in the year 2000, the Far Ultraviolet Spectrograph Explorer (FUSE), will provide very sensitive high-resolution spectra. Present plans call for a resolution of $\lambda/\Delta\lambda \approx 30{,}000$ in the 910–1250Å region to be obtained with a 1.6-m Rowland Spectrograph with about 50 $cm^2$ effective area. There will also be a "planetary" channel covering the 400–1600Å range with a resolution of 500. Cash (1984) and Moos and Friedman (1991) have described earlier versions of the FUSE optical system.

The FUV contains such important diagnostics as the Lyman lines of atomic H and D, electronic bands of the molecules $H_2$ and HD, the resonance lines of C III and O VI, and forbidden lines of coronal ions. Feldman and Doschek (1991) have compiled a very useful list of FUV emission lines observed in the solar spectrum, and have also listed forbidden lines of abundant elements that FUSE should observe in the FUV spectra of stars with hot coronal plasmas. In the EUV, FUSE will have a glancing-incidence spectrograph covering the 100–350Å range with resolutions of approximately 800 and 2200. The EUV contains many bright lines of highly ionized species such as Fe XI–XXIV, which provide useful diagnostics for $10^6$–$10^7$ K plasmas. For a review of the EUV spectra anticipated from hot astrophysical plasmas and useful spectroscopic diagnostic techniques, see, for example, Feldman (1981), Doschek and Cowan (1984), and Doschek (1991).

## 3.3 Spectroscopic Data Needed by the FUV and EUV Instruments of the 1990s

The analysis of FUV and EUV spectra from HUT, FUSE, and the other space instruments described in Table 1 will require new laboratory measurements and theoretical calculations in several areas. Also, it is important to compile accurate on-line data bases for ultraviolet wavelengths, oscillator strengths, and other atomic and molecular parameters. Data bases for Fe I and Fe II in particular are needed.

### 3.3.1 Accurate wavelengths and energy levels

Accurate absolute wavelengths are needed, for example, to measure downflow and wind expansion velocities in stars, particularly in the late-type stars for which the wind expansion velocities in the atmospheric layers accessible to FUV and EUV instruments and the downflow velocities in magnetic flux tubes are expected to be small. Typical values for downflow velocities in dwarf stars are +6 km $s^{-1}$ (Ayres, Jensen, and Engvold 1988). Accurate wavelengths are also needed to Doppler image bright active regions and dark starspots on active late-type stars and regions of high chemical abundances on the surfaces of chemically peculiar B- and A-type stars. The most accurate wavelengths, however, are needed for studies of the velocities and line profiles of interstellar clouds, since the observed lines are very narrow and individual cloud velocities can differ from other clouds on a given line of sight by a few km $s^{-1}$ or less. Since line center wavelengths can be measured to better than 1/10 of the instrumental resolution for strong lines with good signal/noise, laboratory measurements should be at least this accurate. For the EUV spectral region the ORFEUS instrument sets the desired wavelength accuracy of 4 km $s^{-1}$, and for the FUV spectral region the IMAPS instrument sets a far more stringent requirement of 0.2 km $s^{-1}$.

Resonance, intersystem, and forbidden lines of moderately ionized C, N, and O are most important for the analysis of stellar spectra; neutral and singly-ionized C, N, and O, as well as molecules like $H_2$ and HD, are important for interstellar work. Studies of abundances in the interstellar medium generally rely on lines with little saturation, since these lines are favorably placed in curves of growth. For this purpose, lines of less abundant neutral and singly-ionized elements are most useful; in particular, Mg, Si, P, S, A, Cr, Mn, Co, Ni, and Zn. Typically the wavelength accuracy is poorer for the higher stages of ionization. This can make line identifications uncertain in the FUV and EUV, since the spectral resolution and wavelength accuracy of the new generation of instruments are far better than for previous instruments.

### 3.3.2 Oscillator strengths

The uncertainties of abundance analyses for the interstellar medium and stars are typically limited by the accuracy of the oscillator strengths of such important species as C I and II, N I and II, O I and II, Mg I and II, Si II (cf. Shull, Snow, and York 1981), P II, and Fe I and II (cf. Shull, Van Steenberg, and Seab 1983). Among the noble gasses, only argon has its neutral resonance lines located in the FUV, where they may be used to measure the interstellar abundance of this element. Of particular importance to interstellar work at longer wavelengths are the intersystem lines

O I] 1355Å, C II] 2325Å, and Si II] 2335Å, because these weak lines are critical to abundance measurements for these important elements. For higher stages of ionization, a few accurate laboratory measurements are needed to test the accuracy of theoretical predictions for a large number of lines. Oscillator strengths for the FUV lines of N III, S IV and VI, and P IV and V are needed to study the absorption by hot gas in the interstellar medium.

### 3.3.3 Molecular data in the FUV

Accurate wavelengths, oscillator strengths, and photoionization cross sections are needed for six molecules with important electronic spectra in the FUV – $H_2$, HD, CO, $N_2$, $C_2$, and OH. At present the oscillator strengths are in good shape for the first four of these molecules and the $H_2$ fluorescence processes are probably well understood. The molecules $H_2$, HD, and CO show rich absorption line spectra in interstellar clouds (e.g. Jenkins *et al.* 1989; Van Dishoeck and Black 1988), and CO is abundant in the atmospheres of K- and M-type giants and supergiants (Ayres, Moos, and Linsky 1981). Recent GHRS spectra of $\alpha$ Orionis show many of the CO fourth positive bands in absorption. The collisional excitation and de-excitation rates for rotational levels within the electronic bands of $H_2$, and the $H_2$ dissociation rates, all due to collisions with H and $H_2$, should be determined more accurately. Oscillator strengths and dissociation rates are also needed for such trace molecules as $N_2$, $C_2$, and OH to test the photochemistry in interstellar clouds.

### 3.3.4 Electron collisional excitation rates

Electron collisional excitation cross sections are probably the single most important class of atomic data needed for the analysis of FUV and EUV spectra. These data are vital for estimating plasma electron densities and temperatures, as well as for computing ionization equilibria and deriving abundances from emission lines formed in the hot plasmas of stars and shocked or photoionized nebulae. Both experimental and theoretical work are needed in the following areas:

- **Intersystem transitions.** Line flux ratios within intersystem multiplets or between intersystem and permitted lines are commonly used as density diagnostics when the collisional excitation and de-excitation rates for the intersystem transitions are known accurately. Feldman (1981) provides a list of many of the density-sensitive line ratios in the ultraviolet that are useful for studying plasmas with densities in the range $10^8$ – $10^{13}$ $cm^{-3}$. In the FUV, density-sensitive lines are present in C III, N III, Ne V and VI, Si IV, and S III and IV (Feldman and Doschek 1991). In the EUV, important species are O VI, Mg VII and VIII, Si IX, S XI, Ca XV-XVII, and Fe IX-XV and XXI (Dere *et al.* 1979; Dere and Mason 1981; Doschek 1991). The accuracy of these density diagnostics depends on the collisional excitation rates for the intersystem transitions, which have a wide range of uncertainty, and the validity of the assumption of steady-state ionization equilibrium. The latter can be a poor approximation when there are flows in the presence of steep thermal gradients as occur in the solar transition region (cf. Dupree, Moore, and Shapiro 1979).

- **Highly ionized species.** More accurate rates are needed for the higher members of many isoelectronic sequences, in particular the Li (e.g. Mg X, Si XII,...), Be, and B sequences. Excitation rates including cascades are also important for the Fe IX–XIX ions, in particular Ne-like Fe XVII. The "opacity project" (e.g. Seaton 1987) is making major progress in this area, but further work through the "Fe project" is needed.

- **S II–IV.** Since recent HUT spectra of the Io Torus show many of these lines in the spectral region 1000–1730Å, collisional excitation rates are needed to support the analysis of these spectra.

- **Low stages of ionization of abundant species.** The ions of interest include neutral or singly-ionized C, N, O, Mg, Si, S, and Fe.

- **Dielectronic recombination cross sections.** These data are needed for ionization equilibria calculations of photoionized gases, especially at low temperatures ($\sim 10^4$ K) where resonances near threshold are most important. Dielectronic recombination is typically the dominant mode of recombination for astrophysical plasmas at temperatures $T \geq 20{,}000$ K. Recent close coupling calculations by Nahar and Pradhan (1991) show factors of 2–10 disagreement between their total recombination rates and the sum of the radiative and dielectronic recombination rates for the astrophysically important species C I, N II, and O III. This may indicate that many ionization equilibria are now poorly understood.

- **Charge exchange.** Charge exchange collisions with neutral H and He typically dominate the effective neutralization rate for most ions when $H/H^+ \geq 10^{-3}$. This is important, for example, in the outer heliosphere, x-ray binaries, and x-ray ionized nebulae. Recombination usually occurs to excited states followed by line emission. Charge exchange rates involving ionized iron are important for ionization balance calculations, both in equilibrium and out of equilibrium, but these rates are presently unknown.

*Acknowledgements:* I would like to thank Webster Cash, H. Warren Moos, and Mike Shull for their generous input. I would especially like to thank Mike Shull for his comments on the manuscript. I also acknowledge discussions with many colleagues who have participated in the planning for FUSE.

## 3.4 References

Ayres, T.R., Jensen, E., and Engvold, O. 1988, *Ap. J. Suppl.* **66**, 51.

Ayres, T.R., Moos, H.W., and Linsky, J.L. 1981, *Ap. J.* **248**, L137.

Blair, W.P. and Gull, T.R. 1990, *Sky & Telescope* **79**, 591.

Blair, W.P. *et al.* 1991, *Ap. J.* **379**, L33.

Bowyer, S. and Malina, R.F. 1991, in "Extreme Ultraviolet Astronomy", ed. R.F. Malina and S. Bowyer (New York: Pergamon Press), p. 397.

Broadfoot, A.L. *et al.* 1977, *Space Sci. Rev.* **21**, 183.

Cash, W.C. Jr. 1984, *Appl. Optics* **23**, 4518.

Cash, W.C. Jr. *et al.* , 1989, *Exper. Astron.* **1**, 123.

Davidsen, A.F., Kimble, R.A., Durrance, S.T., Bowers, C.W., and Long, K.S. 1991, in "Extreme Ultraviolet Astronomy", ed. R.F. Malina and S. Bowyer (New York: Pergamon Press), p. 427.

Dere, K.P. and Mason, H.E. 1981, in *Solar Active Regions*, ed. F.Q. Orrall (Boulder: Colorado Assoc. Univ. Press), p. 129.

Dere, K.P., Mason, H.E., Widing, K.C., and Bhatia, A.K. 1979, *Ap. J. Suppl.* **40**, 341.

Doschek, G.A. 1991, in "Extreme Ultraviolet Astronomy", ed. R.F. Malina and S. Bowyer (New York: Pergamon Press), p. 94.

Doschek, G.A. and Cowan, R.D. 1984, *Ap. J. Suppl.* **56**, 67.

Dupree, A.K., Moore, R.T., and Shapiro, P.R. 1979, *Ap. J.* **229**, L101.

Feldman, U. 1981, *Physica Scripta* **24**, p. 681.

Feldman, U. and Doschek, G.A. 1991, *Ap. J. Suppl.* **75**, 925.

Feldman, P.D. *et al.* 1991, *Ap. J.* **379**, L37.

Green, J.C., Cash, W., Cook, T.A., and Stern, S.A. 1991, *Science* **251**, 408.

Holberg, J.B. 1991, in "Extreme Ultraviolet Astronomy", ed. R.F. Malina and S. Bowyer (New York: Pergamon Press), p. 8.

Hurwitz, M. and Bowyer, S. 1991, in "Extreme Ultraviolet Astronomy", ed. R.F. Malina and S. Bowyer (New York: Pergamon Press), p. 442.

Jenkins, E.B., Lees, J.F., van Dishoeck, E.F., and Wilcots, E.M. 1989, *Ap. J.* **343**, 785.

Moos, W. and Friedman, S.D. 1991, in "Extreme Ultraviolet Astronomy", ed. R.F. Malina and S. Bowyer (New York: Pergamon Press), p. 457.

Nahar, S. and Pradhan, A.K. 1991, submitted to *Phys. Rev. A*.

Seaton, M.J. 1987, *J. Phys. B* **20**, 6363.

Shull, J.M., Snow, T.P. Jr., and York, D.G. 1981, *Ap. J.* **246**, 549.

Shull, J.M., Van Steenberg, M., and Seab, C.G. 1983, *Ap. J.* **271**, 408.
Spitzer, L., Jr. and Jenkins, E.B. 1975, *Ann. Rev. Astron. Astrophys.* **13**, 133.
Van Dishoeck, E.F. and Black, J.H. 1988, *Ap. J.* **334**, 771.

# 4

# Atomic Data Needed for Analysis of EUV and X-Ray Spectra

**John C. Raymond**[1]

ABSTRACT Given high resolution X-ray spectra, one can derive the temperature structure, elemental abundances, density, and in some cases heating or cooling rate of a hot astrophysical plasma. The accuracy of various atomic rate coefficients limits the accuracy of the inferred plasma parameters. The important rates are typically collisional ionization, radiative and dielectronic recombination, collisional excitation, and radiative decay. Of these, dielectronic recombination is often among the most uncertain. Accurate collisional excitation rates are available for strong transitions of simple ions, but a disturbing number of excitation rates are only known to a factor of two.

## 4.1 INTRODUCTION

Emission lines dominate the X-ray spectra of hot, optically thin astrophysical plasmas. The line intensities can provide powerful diagnostics for the temperature and density distributions of the emitting gas, its elemental composition, and in some cases its rate of heating or cooling. The accuracy with which these physical parameters can be derived depends on the accuracy of the intensity measurements and on the accuracy of the atomic rates used in the interpretation. The interpretation of many observations by the AXAF (Advanced X-ray Astronomy Facility), EUVE (Extreme Ultraviolet Explorer), and XMM (High Throughput X-ray Spectroscopy Mission) satellites will be limited by the atomic rate uncertainties unless great progress is made in the next few years. The following sections summarize the expected spectroscopic capabilities of these satellites and the relevant atomic processes. Mewe (1991) gives an excellent review of thermal X-ray emission models. This paper will concentrate on the uncertainties in the atomic rates.

[1] Harvard-Smithsonian Center for Astrophysics

## 4.2 OBSERVATIONAL PROSPECTS

Good spectral resolution X-ray observations of objects beyond the solar system have only recently become possible. The Solid State Spectrometer (SSS), with about 160 eV resolution, and Focal Plane Crystal Spectrometer aboard the *Einstein* satellite and the transmission gratings on *Einstein* and EXOSAT have provided exciting observations of supernova remnants, stellar coronae, and hot gas in a galaxy cluster between 0.1 and 4 keV (e.g. Mewe 1991; Canizares 1987). The BBXRT experiment on the Space Shuttle extended moderate resolution spectroscopy to the iron K$\alpha$ feature at 6.7 keV, and ASTRO-D will soon provide similar observations of many more objects.

The AXAF and XMM satellites will provide far greater capabilities. Both will have much larger collecting areas, measuring many lines with good statistical accuracy. AXAF will include a CCD imager, providing modest spectral resolution and high spatial resolution over extended areas (Nousek et al. 1987). The expected 120 eV resolution is about 25% better than that of the *Einstein* SSS. This will be excellent for studies of supernova remnants and galaxy clusters. High spectral resolution will be provided by the X-ray calorimeter which is being developed at Goddard Space Flight Center and the University of Wisconsin (Holt 1987). It will measure individual photon energies to better than 10 eV. High spectral resolution will be provided by the "Crystal Spectrometer", which will attain a resolving power up to 2000 (Canizares et al. 1987). A transmission grating will also be provided for observing point sources. It is expected to reach a resolving power of about 1000 near 100 Å (Brinkmann et al. 1987) with an effective collecting area above 10 cm$^2$ over most of the AXAF energy range. ESA's High Throughput X-ray Spectroscopy Mission (XMM) is specifically designed for spectroscopy (see ESA SP-1097). Its CCD detectors will provide modest spectral resolution in themselves (15 at 1.5 keV and 50 at 6 keV), and its reflection gratings will provide resolving power of a few hundred in the 1 keV region with an effective collecting area peaking at a few hundred square centimeters.

## 4.3 EMISSION LINE DIAGNOSTICS

Several types of diagnostics are important. To derive the temperature distribution (i.e., emission measure as a function of temperature) or the relative elemental abundances, one generally needs to know the emissivities of a few of the strongest lines of the most abundant ions. This requires knowledge of the ionization state of each element and the collisional excitation rates. The resonance lines of He-like ions are widely used. These ions dominate over broad temperature ranges, and they are therefore intense and relatively insensitive to uncertainties in the ionization and recombination rates. The trade-off is a loss of resolution in the inferred temperature distribution. The collision strengths of the He-like resonance lines are believed to be accurate to about 10%, both because they have been extensively studied and because of the simplicity of the ions (e.g. Tayal and Kingston 1985).

Density diagnostics involve ratios of two lines of a single ion, one of which involves a metastable level. Examples at X-ray wavelengths include the forbidden-to-resonance line ratios of He-like ions (e.g. Pradhan and Shull 1981) and iron lines excited from different states in the ground configuration (e.g. Mason et al. 1984). These diagnostics involve at least two excitation rates and an intercombination or forbidden transition radiative decay rate. One of the excitation rates is likely to include a strong resonance contribution, so simple theoretical calculations will not suffice. In many cases, the intensity ratio only changes by a factor of 3 to 5 over 2 2 to 3 orders of magnitude density range. The combination of this modest sensitivity with the accumulated uncertainties in several atomic rates typically limits the density diagnostic to about a factor of two accuracy.

A somewhat less common type of diagnostic involves the departure from ionization equilibrium, which would indicate a rapidly heating or cooling plasma. Measurements of the ratio of concentrations of two ions of one element provide such a diagnostic if the temperature is known. It is also possible to measure the temperature from an intensity ratio of two lines of an ion which have differing Boltzmann factors or differing contributions from recombination (Canizares 1987; Doyle et al. 1985). This temperature can then be compared with that expected in ionization equilibrium. Thus the inference of departure from ionization equilibrium involves at least two collisional excitation rates and several ionization and recombination rates. While the temperature sensitivity could be very great if one examines lines having very different excitation potentials, there are few instruments which simultaneously measure lines at very different wavelengths. An instrument which covers a broad wavelength range with good radiometric accuracy should be a high priority for the future.

## 4.4 IONIZATION STATE

### 4.4.1 Ionization Rates

Ionization rates have been computed in the Coulomb-Born and distorted wave approximations (e.g. Younger 1983; Pindzola et al. 1986). Cross sections have been measured in crossed beam experiments for quite a few ions. The compilation by Arnaud and Rothenflug (1985) has recently been assessed (Kato, Masai and Arnaud 1991; Arnaud and Raymond 1992). The major changes in the last five years are the measurements of many iron ions (Gregory et al. 1986, 1987). For the complex low ionization stages, the new rates are several times larger than the earlier ones.

### 4.4.2 Recombination Rates

Radiative and dielectronic recombination contribute to the total recombination rates. The radiative recombination rate is generally computed by adding recombination to the ground level (derived from the photoionization cross section and detailed balance) to a hydrogenic treatment of recombination to excited states. For iron, a more detailed treatment of the excited levels makes little difference to the rates for relatively highly ionized species (Arnaud and Raymond 1991), the largest difference being about 15%. Laboratory measurements for several ions also seem to show that the hydrogenic approximation is fairly good when many levels are included (Andersen and Bolko 1990).

The dielectronic recombination rates are more problematic. While there is good agreement among various calculations for simple cases, such as the Li-like 2s-2p and He-like $1s^2$ - 1s2p transitions, there are substantial disagreements among competing calculations for more difficult cases. Laboratory measurements have yet to confirm the calculations in the weak field limit even for the simple cases. No complete set of dielectronic recombination rates has been published since the work of Jacobs et al. (1977 and subsequent papers), but those rates are inaccurate for high Z (Smith et al. 1985). Figure 4.1 shows various computed rates for Fe XVII. The discrepancy among them is about 40%, and this is typical of the rates for transitions involving a change in principal quantum number. All these rates are 3-4 times larger than those of Jacobs et al. (1977). Arnaud and Raymond (1991) have evaluated the available rates for iron.

Still more problems plague the dielectronic recombination rates. The rates are sensitive to density under conditions found in the solar transition region and flares. Summers (1974) presented a complete set of rates for elements through calcium, but this was before the importance of autoionization to excited levels was appreciated (Jacobs et al. 1977). More recently, Reisenfeld et al. (1991) have pointed out that plasma microfields will partially counteract the collisional effects discussed by Summers, but the rates are still quite uncertain.

### 4.4.3 Ionization Balance

Ionization rate uncertainties are typically 10 - 20%, while dielectronic recombination rate uncertainties may reach a factor of two. Figure 4.2 shows the concentrations of Fe XX according to Arnaud and Raymond (1991) as compared with the Arnaud and Rothenflug (1985) calculation and the current version of the Raymond and Smith (1977) code. All three agree

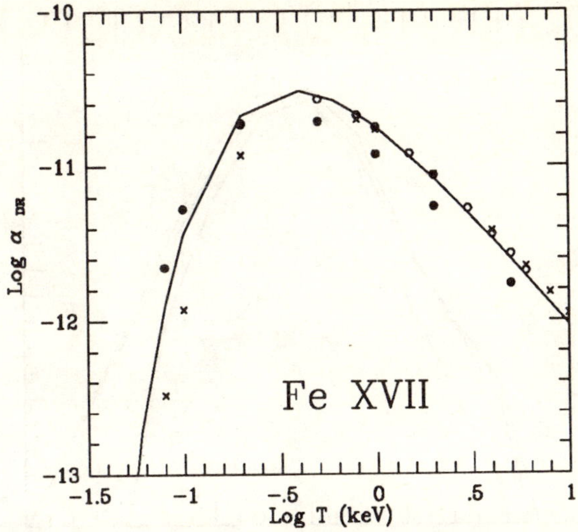


FIGURE 4.1. Dielectronic recombination rates from Chen (open circles), Hahn (filled circles), and Romanik (Xs) with values adopted by Arnaud and Raymond (1991).

near the concentration peak, but the new calculations differ from both of the old ones by substantial factors away from the peak.

## 4.5 COLLISIONAL EXCITATION

The best collision strength calculations are probably good to about 10%, and laboratory measurements confirm them for a few simple cases, particularly $\Delta n = 0$ transitions in Li-like ions. However, many of the strong X-ray lines involve complicated ions, and the data situation is much worse. Calculations including cascades and resonances in the collision cross sections have been performed for He-like and Ne-like ions (Tayal and Kingston 1985; Smith et al. 1985; Goldstein et al. 1989), but they are not available for the $\Delta n \neq 0$ transitions of Li-like through F-like iron which are strong in spectra of stellar coronae and supernova remnants. There is evidence that errors as large as a factor of two are common for individual transitions in these cases. McKenzie et al. (1985) found such discrepancies between predicted and observed relative intensities of lines from the 3s, 3p and 3d levels of Fe XXIV. Among the other reported discrepancies are the singlet to triplet ratios in Fe XV (Dufton et al. 1990).

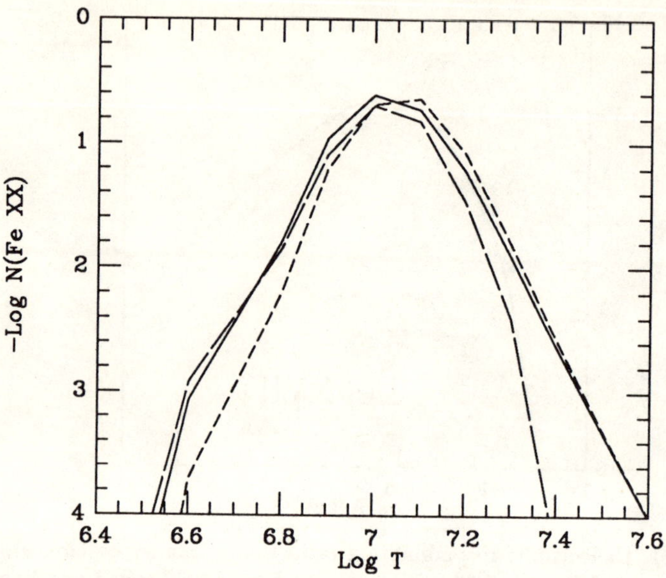


FIGURE 4.2. Fe XX concentrations from Arnaud and Raymond (solid), Arnaud and Rothenflug (dashed) and the 1990 version of Raymond and Smith (long-dashed).

## 4.6 OVERALL UNCERTAINTY

There is no single number describing the uncertainties in the predicted intensities of all X-ray lines. The strong resonance lines of He-like (and to a lesser extent Ne-like and H-like) ions are insensitive to uncertainties in the ionization and recombination rates. Close-coupling collision strengths are generally available for these ions, and the overall uncertainty is only about 10%. The 2s-3p and 2p-3s,3d transitions of Li-like through F-like iron dominate the spectrum near 10 Å, and those of Si and S dominate the 40 - 80 Å band. These lines are sensitive to the ionization balance, and the availably collision strengths are generally Distorted Wave calculations at a few energies, with no account taken of resonance in the cross sections (e.g. Mason and Storey 1980; Mann 1983). The overall uncertainty for these lines is probably around 30% for the strong 2p-3d lines and worse for the 2p-3s lines which are affected by resonances and cascades from higher levels. More complicated ions involving n=3 ground states (Fe IX-XVI) are likely to be still worse (e.g. Dufton et al. 1990). These uncertainties will seriously hamper the interpretation of data from ASTRO-D, AXAF and XMM unless laboratory and theoretical efforts improve the situation soon.

This work was supported by NASA Grant NAGW-528 to the Smithsonian Institution.

# 4.7 REFERENCES


Andersen, L.H., and Bolko, J. 1990, J. Phys. B. 23, 3167.

Arnaud, M., and Rothenflug, R. 1985, Astr. Ap. Suppl., 60, 425.

Arnaud, M., and Raymond, J. 1991, submitted to Ap. J.

Brinkmann, A.C., van Rooijen, J.J., Bleeker, J.A.M., Dijkstra, J.H., Heise, J., De Korte, P.A.J., Mewe, R., and Paerels, F. 1987, Astrophys. Lett. and Comm, 26, 73.

Chen, M.H. 1986, Phys. Rev. A, 34, 1073.

Canizares, C.R. 1990, in *High Resolution X-ray Spectroscopy of Cosmic Plasmas*,P. Gorenstein and M. Zombeck, eds. (Cambridge U. Press: Cambridge), p. 136.

Canizares, C.R., et al. 1987, Astrophys. Lett. and Comm., 26, 87.

Doyle, J.G., Raymond, J.C., Noyes, R.W., and Kingston, A.E. 1985, Ap. J., 297, 816.

Dufton, P.L., Kingston, A.E., and Widing, K.G. 1990, Ap. J., 353, 322.

Goldstein, W.H., Osterheld, A., Oreg, J., and Bar-Shalom, A. 1989, Ap. J.,Lett., 344, L37.

Gregory, D.C., Meyer, F.W., Muller, A., and Defrance, P. 1986, Phys. Rev. A, 34, 3657.

Gregory, D.C., Wang, L.J., Meyer, F.M., and Rinn, K. 1987, Phys. Rev. A, 35, 3256.

Hahn, Y. 1989, JQSRT, 41, 315.

Holt, S.S. 1987, Astrophys. Lett. and Comm., 26, 61.

Jacobs, V.L., Davis, J., Kepple, P.C., and Blaha, M. 1977, Ap. J., 211, 605.

Kato, T., Masai, K., and Arnaud, M. 1991, National Institute for Fusion Science Research Report NIFS-DATA-14.

Mann, J.B. 1983, At. Data and Nucl. Data Tables, 29, 407.

Mason, H.E., and Storey, P.J. 1980, MNRAS, 191, 631.

Mason, H.E., Bhatia, A.K., Kastner, S.O., Neupert, W.M., and Swartz, M. 1984, Solar Phys., 92, 199.

McKenzie, D.L., Landecker, P.B., Feldman, U., and Doschek, G.A. 1985, Ap. J., 289, 849.

Mewe, R. 1991, Astr. Ap. Review, 3, 127.

Nousek, J.A., Garmire, G.P., Ricker, G.R., Collins, S.A., and Reigler, G.R. 1987, Astrophys. Lett. and Comm., 26, 35.

Pindzola, M.S., Griffin, D.C., and Bottcher, C. 1986, Phys. Rev. A, 34, 3368.

Pradhan, A.K., and Shull, J.M. 1981, Ap. J., 249, 821.

Raymond,J.C., and Smith, B.W. 1977, Ap. J. Suppl., 306, 762.

Reisenfeld, D. B., Raymond, J.C., Young, A.R., and Kohl, J.L. 1991, submitted to Ap.J. Letters.

Romanik, C.J. 1988, Ap. J., 330, 1022.

Smith, B.W., Raymond, J.C., Mann, J.B., Jr., and Cowan, R.D. 1985, Ap. J., 298, 898.

Summers, H.P. 1974, Culham Laboratory Internal Memo IM-367, Culham Laboratory, Ditton Park, Slough, England.
Tayal, S.S., and Kingston, A.E. 1985, J. Phys. B, 18, 2983.
Younger, S.M. 1983, JQSRT, 29, 67.

# 5

# Atomic and Molecular Data for Observations of the Interstellar Medium with the Hubble Space Telescope

**Donald C. Morton**[1]

ABSTRACT The high resolution spectrograph on HST, with resolving powers up to 94000 or 3.2 km s$^{-1}$, is giving us a new look at interstellar absorption lines in the range 1090 to 3350 Å. The high signal-to-noise ratios possible with the digicon detectors provide precise line profiles suitable for studying individual velocity components.

Comparison of these components in different atoms and molecules requires accurate laboratory wavelengths. Their uncertainties are within ±0.002Å for many of the transitions expected to be detected by HST and almost all are known within ±0.02 Å. The notable exceptions are some intersystem transitions, which are mentioned in this paper.

Transition probabilities with 1σ errors within ±0.03 dex and good agreement between theoretical and experimental values are becoming available for many of the resonance lines observable with HST, but many remain to be measured to this accuracy, and some important lines are not yet known within ±0.3 dex. This paper lists those values urgently needed to take full advantage of HST.

## Introduction

The ultraviolet wavelengths from 1090 to 3350 Å, which are observable with the Goddard High Resolution Spectrograph (GHRS) on the Hubble Space Telescope (HST), are particularly important for studies of the interstellar gas. This range includes many of the resonance lines of the more abundant species such as H I, C I, C II, O I, Mg II, Si II and Fe II in the gas between

---

[1]Herzberg Institute of Astrophysics, National Research Council of Canada, 100 Sussex Drive, Ottawa, Ontario, Canada K1A 0R6

the stars. There the low particle density and weak radiation field are insufficient to excite many atoms or ions much above the ground state. Essentially all the absorptions occur from the ground levels or from the lowest (< 300 $cm^{-1}$) excited fine-structure levels of the ground term. The resonance lines of Li I, Na I, K I, Ca I, Ca II, and Fe I are observable through the earth's atmosphere, but for most of the abundant atoms and ions the first excited level with different parity from the ground lies above 33000 $cm^{-1}$ so that their resonance lines occur at wavelengths $\lambda < 3000$ Å.

This large separation of the lowest levels also means that many of the strongest permitted lines in the spectra of planetary nebulae and diffuse nebulae occur at ultraviolet wavelengths.

The far UV line spectra of stars were observed first with rockets (Morton and Spitzer 1966). Stone and Morton (1967) identified possible interstellar absorption lines of C II, O I, Al II and Si II, but the equivalent widths were rather large, probably because of blending with stellar lines. The *Copernicus* satellite (Rogerson et al 1973) provided many high resolution (12 km $s^{-1}$ FWHM) UV spectra with good counting statistics over most of the range from the Lyman limit to the atmospheric cut off. The International Ultraviolet Explorer (Boggess *et al* 1978) was able to observe much fainter stars longward of about 1150 Å, but at the lower resolution of 25 km $s^{-1}$ and generally poorer signal-to-noise ratios. Recently Jenkins *et al* (1989) have obtained a far UV spectrum of $\pi$ Sco from 1003 to 1172 Å at a resolution of 2.3 km $s^{-1}$ FWHM with a rocket-borne echelle spectrograph.

Now with HST operating in orbit, we have a new source of high quality spectroscopic data (Savage *et al* 1991, Cardelli *et al* 1991, Smith *et al* 1991). The 0.25 x 0.25 $arcsec^2$ small science aperture used with either of the echelle modes of the GHRS has an instrumental profile of 3.2 to 3.8 km $s^{-1}$ FWHM. See Fig. 1. The actual ranges are 1090 to 1700 Å for Echelle A and 1700 to 3100 Å for Echelle B, with much reduced sensitivity below 1150 Å. Unfortunately the degraded telescope image reduces the fraction of light passing through the aperture by a factor of 4.5 over the prelaunch expectations. The 2 x 2 $arcsec^2$ large science aperture has about 50 to 80% of the expected efficiency, but loses a factor 2 in resolution. Certainly the sorting of interstellar lines which overlap in velocity is done best with the small aperture.

About the same time as HST obtained its first high-resolution spectra, I was able to complete a compilation of wavelengths, transition probabilities and radiation damping constants of resonance atomic lines longward of the Lyman limit. To the extent possible with the available laboratory and theoretical data, the goal was to include the transitions of all elements from H to Ge that might be detected as interstellar absorption lines with the GHRS.

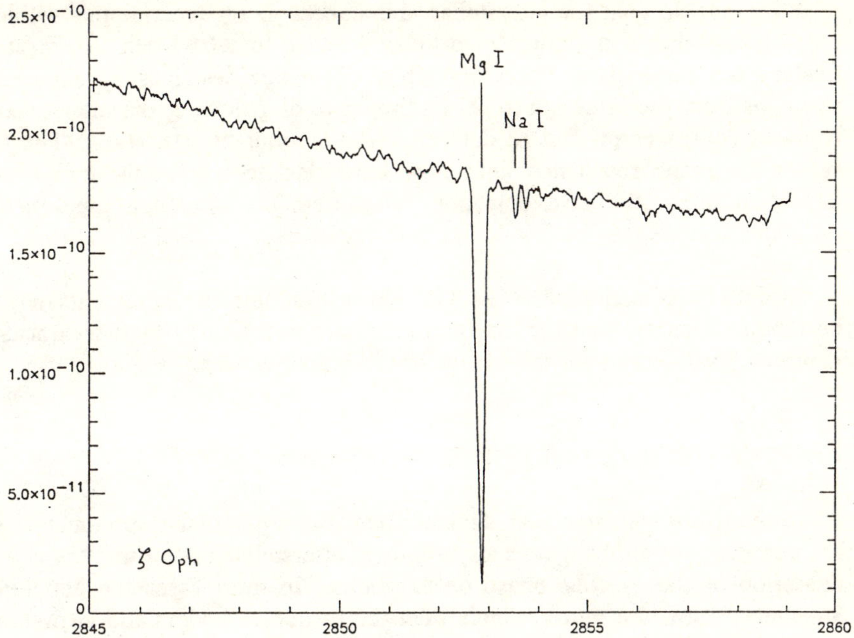


Figure 1. Interstellar absorption lines of Mg I and Na I towards ζ Oph observed with Echelle B on the high resolution spectrograph on HST.

The reader should consult that publication (Morton 1991) for further details about the atomic transitions mentioned here.

# Data Accuracy

The interpretation of HST observations requires accurate laboratory wavelengths for the identification of lines, the determination of radial velocities and the matching of components in the profiles of different transitions. The centres of symmetric lines should be measurable to 1/10 or less of the line width so that the laboratory data must be significantly more accurate than 0.3 km $s^{-1}$ or 1 part in $10^6$. The absolute accuracy of HST wavelengths has not yet reached this level because of thermally and magnetically induced drifts. Cardelli *et al* (1991) found shifts up to about $\pm 2$ km $s^{-1}$ between different observations of ξ Per, but were able to reduce these to $\pm 0.5$ km $s^{-1}$ by comparison of lines at the same ion at different wavelength settings and the use of telluric O I. Further improvements are expected.

A reasonable goal for interstellar abundances is 1$\sigma$ uncertainties within ±0.1 dex so that it is desirable to know *f*-values to ±0.03 dex. Accurate *f*-values are particularly important when one combines lines of different strengths from the same ion to obtain the curve of growth of the interstellar cloud or compares profiles to determine the amount of saturation. The *f*-values for many resonance lines now have this accuracy, but numerous important ones still have only poor measurements or rough theoretical estimates available, and a few do not have even these.

Cardelli *et al* comment: "The UV absorption line measurements have become so accurate that the interpretations are now limited by inaccuracies in atomic *f*-values and our knowledge of solar abundances."

## Solar-System Abundances

Comparisons with standard element abundances for solar-system material are necessary for studying nucleosynthesis in interstellar matter and the condensation of the gaseous phase on to grains. In most cases the adopted abundances now are more reliable than the *f*-values. Anders and Grevesse (1989) have reviewed the meteorite data, along with analyses of the solar photosphere for C, N and O, and quoted 1$\sigma$ errors of ±0.03 dex or less for 15 of the 32 elements from He to Ge, and within ±0.06 for all but Ne and Ar. These last two depended on observations of nearby nebulae and hot stars. Recent determinations are close to those adopted by Anders and Grevesse. In the solar photosphere Biémont *et al* (1990) and Grevesse *et al* (1990) found the abundance of nitrogen lower by 0.06 and 0.05 dex respectively, and Biémont *et al* (1991a) found oxygen lower by 0.07 dex. From an analysis of B stars Keenan (1990) reported argon lower by 0.07 dex.

## Wavelengths

Fortunately, many of the analyses of laboratory spectra now provide more than adequate accuracy for the investigation of the best possible HST data. In most modern work the authors have calculated wavelengths from the energy levels, which have been derived from the measurements of many different transitions, and not just the resonance line in question. A later section will list some of the cases where improved wavelengths are needed.

Here I would like to insert a plea to use vacuum wavelengths in all analyses of astronomical spectra. Vacuum values come naturally from the more fundamental tables of energy levels, and they are essential shortward of 2000 Å. Infrared researchers already work with vacuum wavenumbers and radio astronomers use frequencies. Only in the narrow band from 2000 to

10000 Å is the conversion made to air wavelengths. If vacuum data are used for the calibration spectra, the observed spectra automatically will be in vacuum wavelengths, and there will be no question whether the correct conversion formula to air was used. One by-product would be the avoidance of systematic errors in QSO redshifts when observed air wavelengths are divided by laboratory vacuum values. Certainly an immediate goal should be to avoid the discontinuity in HST wavelengths that would result from quoting air values longward of 2000 Å.

The staff of the U.S. National Institute of Standards and Technology and their collaborators, starting with Charlotte Moore's work in 1950, have contributed much to astrophysics by their critical compilations of atomic energy levels. Publications in the *National Standard Reference Data Series* and in the *Journal of Physical and Chemical Reference Data* now have covered all ions of all elements from H to Cu except Li, Be, B II-V, O II, F, Ne, Cl, and Ar. Some of the important lines of these remaining elements are included in the useful compilation of wavelength standards by Kaufman and Edlén (1974), and other more recent determinations are listed by Morton (1991). Most UV wavelengths now are quoted to 0.001 Å implying uncertainties within ±0.02 Å and many have four decimal places with maximum uncertainties of ±0.002 Å. The lines listed by Kaufman and Edlén usually have even better accuracy.

## Transition Probabilities

During the past decade there has been much progress in the determination of atomic lifetimes, transition probabilities and oscillator strengths (*f*-values). Wiese (1987) has summarized many of the encouraging developments in experiments and theoretical calculations. I would like to mention the following which have particular relevance to studies of the interstellar gas.

### *Laser-excited Lifetimes*

Selective excitation of a level by a tuned laser provides decays without cascades from other levels. Gaupp, Kuske and Andrä (1982) used this technique to obtain accurate lifetimes of the 3p $^2P^0{}_{1/2}$ levels of Li I and Na I. Unfortunately there is some question whether their $1\sigma$ errors are as small as their quoted 0.15 and 0.18% respectively, because of a small systematic deviation from the best theories. The maximum correction to Li I would be a 2.4% increase on $f$ (Wiese 1987). The precise ratio $f(D_1)/f(D_2)$ by Gawlik *et al* (1979) gave the *f*-value for the other Na I line without the usual assumption of LS coupling and showed that the deviation from the predicted line ratio was only 0.9%.

Laser excitation has produced lifetimes with errors of 1% or less for the 3p $^2P^0_{3/2,1/2}$ levels of Mg II (Ansbacher, Li, and Pinnington 1989) and the 4p $^2P^0_{3/2,1/2}$ levels of Ca II (Gosselin, Pinnington and Ansbacher 1988b). The theoretical calculations of Theodosiou and Curtis (1988) and Theodosiou (1989) reproduced the experimental values within their 1$\sigma$ errors. In the case of Ca II a correction for the branch to 3d $^2D$ was necessary in order to derive the *f*-value. Similar lifetimes are now available for Ti II (Gosselin, Pinnington and Ansbacher 1988a) and Fe II (Biémont *et al* 1991b) but more information on branching ratios is needed for these complex species before accurate *f*-values can be derived.

This technique also was used by Kwong, Smith, and Parkinson (1982) to measure the $3s^2\ ^1S$ - $3p\ ^3P^0$ transition of Mg I at 4572 Å. Although $f = 2 \times 10^{-6}$, it is worth searching for this intersystem line with ground-based telescopes.

## *Measurement of Many Transition Probabilities With Moderate Accuracy*

A few laboratories now are producing large numbers of *f*-values and damping constants through partly automated procedures. For example O'Brian *et al* (1991) combined their measurements of the lifetimes of 186 levels of Fe I by laser-induced fluorescence with branching ratios derived from emission-line intensities. Then they extended their list by interpolating between the level populations in the plasma source for a total of 1814 transition probabilities, including many involving the $^5D_4$ ground state.

## *Ion Trap For Intersystem Transitions*

The measurement of the long radiative lifetimes of metastable states often is complicated by collisional de-excitations. These can be minimized in the low densities attainable inside the potential well of a radio-frequency ion trap. Since the theoretical calculations of these transition probabilities also are difficult, it is very important to have reliable experimental values. Where the corresponding interstellar absorption lines are detectable, they are useful for the determination of column densities without saturation corrections. Lifetime measurements are available for C III] λ1909 by Smith *et al* (1983), O III] λλ1666, 1661 by Johnson, Smith and Knight (1984), and Si III] λ1892 by Kwong *et al* (1983). Knight (1982) measured N II λλ2144, 2140, but Smith *et al* (1984) noted that collisional deexcitation may have contributed to the measured lifetime.

## Damping Constants

Some interstellar absorption lines are strong enough to produce damping wings. Lyman $\alpha$ invariably shows this effect, it often is present in C II $\lambda$1335, C II$^*$ $\lambda$1336, O I $\lambda$1302, and Mg II $\lambda\lambda$2504, 2796, and it may influence the profile of many other lines. Since the lower level of a resonance line has an infinite natural life, the damping constant $\gamma$ for the line is simply the reciprocal of the lifetime $\tau$ of the upper level. However, $\gamma$ is equivalent to the line's transition probability $A$ only if there are no branches to other levels; otherwise $\gamma$ is the sum of all the $A$-values for the upper level. For the analysis of these more complicated transitions, it is important to know the branching ratios so that $f$ and $\gamma$ are internally consistent.

Towards $\xi$ Per, Cardelli *et al* (1991) compared column densities derived from strong lines with damping wings, and weak lines of the same ion. They found reasonable agreement for C II and O I, but a discrepency in Mg II that most likely is caused by the poor $f$-value for the weak doublet.

## Needs For Improved Wavelengths

C IV: Moore (1970) quotes 1548.202 and 1550.774 Å for this important doublet, and the reciprocals of her energy levels are 1548.187 and 1550.772, while Rottman *et al* (1990) derived 1548.195 and 1550.770 from rocket observations of the sun. Their errors of $\pm$0.008 Å correspond to 1.6 km s$^{-1}$. New laboratory measurements are needed to reduce these uncertainties. Accurate radial velocities are required to help determine the locations of the C IV absorption.

Intersystem Lines: These transitions invariably are weak, but those originating from the more abundant ions may be detectable in absorption, and almost all of them should be found in emission nebulae. There are several intersystem lines for which, to my knowledge, we still have only rough wavelength estimates. In the HST range these are

| | | |
|---|---|---|
| Ne V] | $2p^2\ ^3P$ - $2p^3\ ^5S^0$ | $\lambda\lambda$ 1146.1, 1137.0 |
| S IV] | $3p\ ^2P^0$ - $3p^2\ ^4P$ | $\lambda\lambda$ 1423.83 - 1398.05 |
| Cl IV] | $3p^2\ ^3P$ - $3p^3\ ^5S^0$ | $\lambda\lambda$ 1436, 1419 |
| Cl V] | $3p\ ^2P^0$ - $3p^2\ ^4P$ | $\lambda\lambda$ 1183.3 - 1162.8 |
| Ar V] | $3p^2\ ^3P$ - $3p^3\ ^3S^0$ | $\lambda\lambda$ 1192.3, 1174.7 |

# Needs For Improved Oscillator Strengths

There are a number of transitions where additional measurements or calculations could have a significant impact on our interpretation of interstellar spectra observable with HST.

Na I λλ3304, 3302: This pair of weak lines is essential for the determination of column densities wherever the D lines are saturated. The theoretical transition probabilities for λλ3304, 3303 and a branch to 3d $^2$D imply a lifetime 10% longer than the mean of the measurements. The next Na I doublet at λ2854 is present in the spectrum ζ Oph (Fig. 1) and also needs an experimental check on the *f*-value.

Mg II λλ1241, 1240: This doublet shown in Fig. 2 has an *f*-value ~ $10^2$ times smaller than its Na I counterpart, and is even more important for interstellar column densities because the lines λλ2804, 2796 usually are saturated. However no one has measured the transition probability of the weak doublet, and the theoretical results range over a factor 5 because of a strong cancellation of terms. If we assume that the damping wings of Mg II λ2796 give the correct column density for ξ Per, the analysis by Cardelli *et al* (1991) implies that log λf(1239.925) = -0.98 rather than -0.48 adopted by Morton (1991).

Si II: This ion has 7 resonance lines from 1190 to 2336 Å with a range of 5 in log λf, and 2 more lines that were observable with *Copernicus*. This should be an ideal case for the determination of a curve of growth, but there remain serious inconsistencies between some of the experimental and theoretical *f*-values, and between each of these and the strengths of the interstellar lines. New experiments are needed for all the transitions, and particularly for the weak line at 1808 Å.

P II: There are large uncertainties in the *f*-values of all the resonance lines of this ion. Three lines at 1533, 1302, and 1153 Å are observable easily with HST.

Cr II: The a $^6$S - z $^6$P$^0$ triplet at 2066, 2062, 2056 Å is observed in the interstellar gas and often is used as an indicator of depletion on grains. A good theoretical *f*-value for the multiplet is available from Aashamar and Luke (1990), but an experimental check would be worthwhile.

Mn II: The a $^7$S - y $^7$P$^0$ triplet at 1201, 1199, 1197 Å appears in interstellar spectra, but the adopted *f*-values were derived by a curve-of-growth comparison with the strengths of the triplet at 2606, 2594, 2577 Å. Independent measurements of the y $^7$P$^0$ triplet and the weaker x $^7$P$^0$ triplet at 1164, 1163, 1162 Å are very desirable.

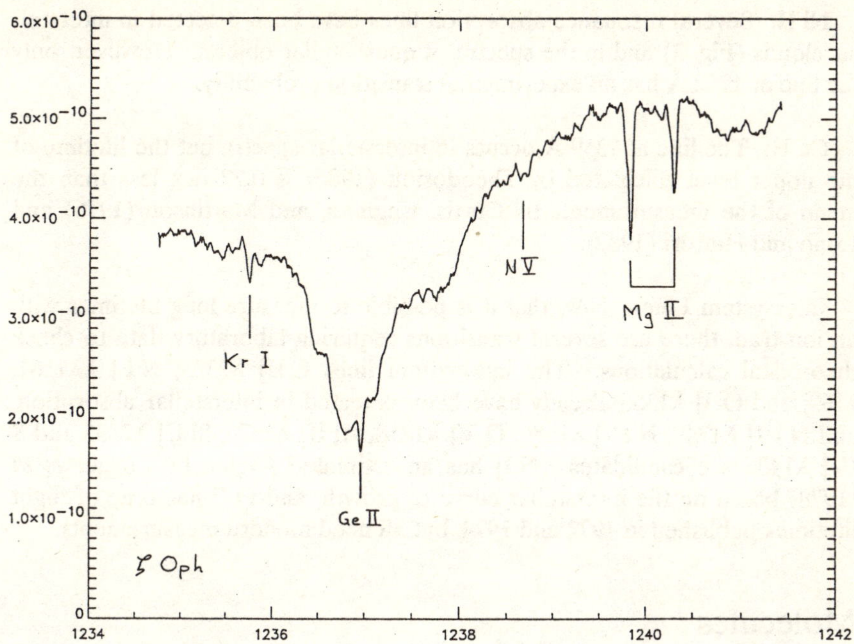


Figure 2. This weak 3s $^2S$ - 4p $^2P^0$ doublet of Mg II towards ζ Oph would give a reliable column density if its *f*-value were better known. This spectrum, observed with Echelle A, includes absorption lines identified with transitions of Kr I, Ge II, and N V.

Fe II: At least 15 resonance lines with a wide range of *f*-values are observable with HST. However, only 4 of these lines have reliable experimental *f*-values. Nussbaumer, Pettini, and Storey (1981) calculated many *f*-values, but comparison with the experimentral data indicates that a correction of about -0.12 dex should be applied to the theoretical log $\lambda f$. Laboratory measurements are needed for some of the weaker short-wavelength lines before the Fe II curve of growth and column density can be obtained reliably. At present we must depend on *f*-values derived from interstellar line strengths.

Co II: This ion has not yet been detected as interstellar absorption, but the strongest lines at 2926, 1941, 1575, and 1466 Å are candidates. Measurements would be useful to check the calculations of Kurucz (1989).

Ni II: Several resonance absorption lines have been detected in interstellar clouds (Fig. 3) and in the spectra of quasi-stellar objects. However, only the line at 1752 Å has an experimental transition probability.

Cu II: The line at 1359 Å occurs in interstellar spectra but the lifetime of the upper level calculated by Theodosiou (1986) is 0.22 dex less than the mean of the measurements by Curtis, Engman, and Martinson (1976) and Kono and Hattori (1982).

Intersystem Lines: Now that it is possible to measure long lifetimes with an ion trap, there are several transitions requiring laboratory data to check theoretical calculations. The intersystem lines C II] λ2325, N I ] λλ1161, 1160, and O I] λ1356 already have been detected in interstellar absorption, and N III] λ1749, N IV] λ1486, O V] λ1218, Al II] λ2670, Si II] λ2335 and S IV] λ1405 are candidates. N I] has an estimated *f*-value by Lugger *et al* (1978) based on the interstellar curve of growth, and O I] has time-of-flight lifetimes published in 1972 and 1974, but all need modern measurements.

## Molecules

The GHRS spectral range includes many molecular electronic transitions from the low rotational levels of the lowest vibrational level, which could occur as interstellar absorption lines. The (6,0) band of the fourth positive system of CO appears in the spectrum of ξ per shown in Fig. 3. In the (2,0) and (3,0) CO bands at 1478 and 1447 Å toward this star, Smith *et al* (1991) found the R(0), R(1) and Q(1) lines, but none from $J \geq 2$ at 11.53 $cm^{-1}$. A few of the longest wavelength transitions of the $H_2$ Lyman system and the corresponding HD lines should be detectable between 1090 and 1126 Å. HST also will be able to record the absorption lines of CH at 3138, 3144 and 3147 Å known from ground-based observations, the lines of OH at 1222 Å measured by *Copernicus* (Snow 1976), and possibly the $C_2$ line at 1314 Å because lines of this molecule at 7719 and 8758 Å have been found. Furthermore, the models of van Dishoeck and Black (1986) indicate that HCl, $H_2O$ and $CH_2$ should be detectable.

Hsu and Hayden Smith (1977) have published a useful compilation of molecular electronic transitions and Jenkins et al (1973), Morton (1975), and Snow (1975, 1976, 1980) have listed the results of various searches. Table 1 updates Hsu and Hayden Smith, where improved data are now available for the range of the GHRS, and adds two triatomic molecules. Isotopes such as $^{13}C^{16}O$ also will be important. To conserve space, the table lists only the transitions involving the lowest rotational level, though absorptions from higher J″ levels of the lowest vibrational state are possible. When using Hönl-London factors to calculate the *f*-values for multiple absorptions from

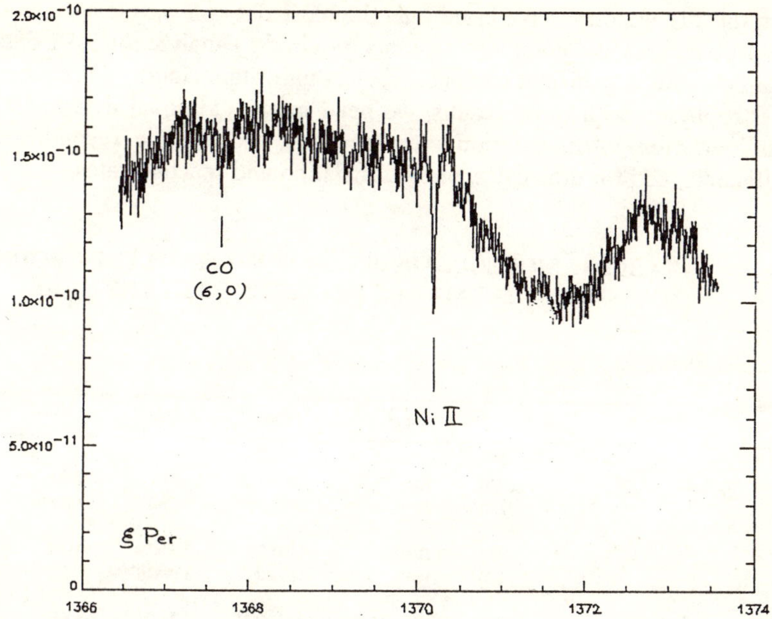


Figure 3. Absorption lines of CO (6,0) and Ni II observed with Echelle A towards ζ Per. The Ni II line does not have a measured *f*-value.

the same rotational level, it is important to use a consistent set of definitions, as described by Whiting *et al* (1980).

The theoretical *f*-values for $H_2$ include the perturbation correction calculated by Ford (1975). The *f*-values derived from the experimental lifetimes of the A $^1\Pi$ - X $^1\Sigma^+$ band of CO by Field *et al* (1983) are in reasonable agreement with the theoretical work of Kirby and Cooper (1989). Their calculations also agree with the experimental values for B $^1\Sigma^+$ (0,0) by Lee and Guest (1981) but are a factor 2 or more below the experimental results of Letzelter *et al* (1987). Most other molecular *f*-values are based on single measurements or calculations, and some, such as MgH+, AlH, SH and CaH that may be detectable with the GHRS, have no UV data.

Herzberg's (1961a) laboratory spectrum of $CH_2$ shows a broad absorption at 1415.5 Å that is assumed to originate from the ground state. This paper also shows two narrow absorptions at 1410.1 and 1396.8 Å that may have the same origin, and Herzberg (1961b) found absorption lines at 1306.3, 1261.9 and 1239.3 Å that form a Rydberg series with 1415.5 Å. The spectra in the former paper include some absorption bands of $CH_3$ that might have ground state origins and hence would be worth looking for as well.

In this discussion, I have listed only the basic data on wavelengths, transition probabilities and damping constants required to analyze the HST data to obtain column densities of each absorber in individual clouds. Of course, to do astrophysics with these results, we need many additional data including photoionization rates, collision cross sections, dielectronic recombination coefficients, and for molecules, photodisociation and reaction rates.

## TABLE 1. MOLECULAR ABSORPTIONS FROM THE GROUND STATE THAT MAY BE DETECTED WITH THE GHRS

**a) Diatomic Molecules**

| Mol. | System | (v'v") | (J") | $\lambda_{vac}$ | $f_{J'J"}$ | Ref. |
|---|---|---|---|---|---|---|
| $H_2$ | $B\,^1\Sigma_u^+ - X\,^1\Sigma_g^+$ | (0,0) | R(0) | 1108.128 | $1.73\times10^{-3}$ | D84 |
| | | (1,0) | R(0) | 1092.195 | $5.96\times10^{-3}$ | MD76 |
| HD | $B\,^1\Sigma_u^+ - X\,^1\Sigma_g^+$ | (0,0) | R(0) | 1105.837 | $7.60\times10^{-4}$ | W68 |
| | | (1,0) | R(0) | 1092.000 | $2.99\times10^{-3}$ | AD70 |
| CH | $C\,^2\Sigma^+ - X\,^2\Pi_r$ | (0,0) | ${}^PQ_{12}(1/2)$ | 3146.907 | $4.3\times10^{-3}$ | L84 |
| | | | $Q_2(1/2)$ | 3144.094 | $4.3\times10^{-3}$ | |
| | | | ${}^QR_{12}(1/2)$ | | $2.1\times10^{-3}$ | |
| | | | $R_2(1/2)$ | 3138.485 | $2.1\times10^{-3}$ | |
| $C_2$ | $F\,^1\Pi_u - X\,^1\Sigma_g^+$ | (0,0) | R(0) | 1341.63 | 0.02 | HLM69 |
| | | (1,0) | R(0) | 1314.17 | | PRSR83 |
| CO | $A\,^1\Pi - X\,^1\Sigma^+$ | (0,0) | R(0) | 1544.451 | $1.56\times10^{-2}$ | SBT69 |
| | | (1,0) | R(0) | 1509.750 | $3.43\times10^{-2}$ | TS72 |
| | | (2,0) | R(0) | 1477.568 | $4.12\times10^{-2}$ | K76 |
| | | (3,0) | R(0) | 1447.355 | $3.61\times10^{-2}$ | FBLLT83 |
| | | (4,0) | R(0) | 1419.046 | $2.58\times10^{-2}$ | KC89 |
| | | (5,0) | R(0) | 1392.525 | $1.61\times10^{-2}$ | |
| | | (6,0) | R(0) | 1367.622 | $9.1\times10^{-3}$ | |
| | | (7,0) | R(0) | 1344.183 | $4.8\times10^{-3}$ | |
| | | (8,0) | R(0) | 1322.147 | $2.4\times10^{-3}$ | |
| | | (9,0) | R(0) | 1301.403 | $1.1\times10^{-3}$ | |
| | | (10,0) | R(0) | 1281.866 | $0.5\times10^{-3}$ | |
| | | (11,0) | R(0) | 1263.433 | | |
| | | (12,0) | R(0) | 1246.059 | | |
| | | (13,0) | R(0) | 1229.668 | | |
| | | (14,0) | R(0) | 1214.227 | | |
| | | (15,0) | R(0) | 1199.678 | | |
| CO | $B\,^1\Sigma^+ - X\,^1\Sigma^+$ | (0,0) | R(0) | 1150.480 | $2.1\times10^{-3}$ | ERLLLR87 |
| | | (1,0) | R(0) | 1123.570 | $3\times10^{-4}$ | KC89 |
| | | (2,0) | R(0) | 1098.998 | $6\times10^{-6}$ | |

# TABLE 1. MOLECULAR ABSORPTIONS FROM THE GROUND STATE THAT MAY BE DETECTED WITH THE GHRS

**a) Diatomic Molecules**

| Mol. | System | (v'v") | (J") | $\lambda_{vac}$ | $f_{J'J"}$ | Ref. |
|---|---|---|---|---|---|---|
| OH | $A\,^2\Sigma^+ - X\,^2\Pi_i$ | (0,0) | $P_1(3/2)$ | 3082.559 | $4.02\times10^{-4}$ | C80 |
| | | | $Q_1(3/2)$ | 3079.328 | $5.57\times10^{-4}$ | WH80 |
| | | | $R_1(3/2)$ | 3072.902 | $1.31\times10^{-4}$ | SA73 |
| | | (1,0) | $P_1(3/2)$ | 2822.532 | $1.1\times10^{-4}$ | LvDWD82 |
| | | | $Q_1(3/2)$ | 2819.956 | $1.6\times10^{-4}$ | |
| | | | $R_1(3/2)$ | 2814.832 | $0.37\times10^{-4}$ | |
| | | (2,0) | $P_1(3/2)$ | 2616.277 | $1.9\times10^{-5}$ | |
| | | | $Q_1(3/2)$ | 2614.180 | $2.6\times10^{-5}$ | |
| | | | $R_1(3/2)$ | 2610.005 | $0.61\times10^{-5}$ | |
| | | (3,0) | $P_1(3/2)$ | 2450.032 | | |
| | | | $Q_1(3/2)$ | 2448.296 | | |
| | | | $R_1(3/2)$ | 2444.841 | | |
| | $D\,^2\Sigma^- - X\,^2\Pi_i$ | (0,0) | $P_1(3/2)$ | 1222.524 | $4\times10^{-3}$ | D74 |
| | | | $Q_1(3/2)$ | 1222.071 | $5\times10^{-3}$ | vDLD83 |
| | | | $R_1(3/2)$ | 1221.166 | $3\times10^{-3}$ | S76 |
| HCl | $C\,^1\Pi - X\,^1\Sigma^+$ | (0,0) | R(0) | 1290.257 | $1.85\times10^{-1}$ | TGV70 |
| | | (1,0) | R(0) | 1247.079 | $2.2\times10^{-2}$ | SYBP80 |
| | | (2,0) | R(0) | 1209.6 | $2.4\times10^{-3}$ | vDvHD82 |
| | | (3,0) | R(0) | 1175.2 | $6.0\times10^{-5}$ | |
| | $D\,^1\Pi - X\,^1\Sigma^+$ | (0,0) | R(0) | 1211.988 | | TG71 |

**b) Triatomic Molecules**

| Mol. | System | (v'v") | (J") | $\lambda_{vac}$ | $f_{J'J"}$ | Ref. |
|---|---|---|---|---|---|---|
| $H_2O$ | $\tilde{C}\,^1B_1 - \tilde{X}\,^1A_1$ | (0,0,0)-(0,0,0) | $1_{10}-0_{00}$ | 1239.728 | 0.010 | J63 |
| | | | | | | vDB86 |
| | $\tilde{E}\,^1B_1 - \tilde{X}\,^1A_1$ | (0,0,0)-(0,0,0) | $1_{10}-0_{00}$ | 1128.500 | | GCJ91 |
| | $\tilde{E}'\,^1B_2 - \tilde{X}\,^1A_1$ | (0,0,0)-(0,0,0) | $1_{01}-0_{00}$ | 1124.481 | | GCJ91 |
| | $\tilde{F}\,^1A_1 - \tilde{X}\,^1A_1$ | (0,0,0)-(0,0,0) | $1_{11}-0_{00}$ | 1114.225 | 0.030 | GCJ91 |
| | | | | | | SYGB81 |
| $CH_2$ | $\tilde{B}\,^3A_2 - \tilde{X}\,^3B_1$ | (0,0,0)-(0,0,0) | R(0) | 1415.5 | | H61a |

**TABLE 2. References For Molecular Data In Table 1**

| | | |
|---|---|---|
| AD70 | Allison and Dalgarno (1970) | HD theoretical *f*-values |
| C80 | Coxon (1980) | OH energy levels |
| D84 | Dabrowski (1984) | $H_2$ energy levels |
| D74 | Douglas (1974) | OH energy levels |
| ERLLLR87 | Eidelsberg, Roncin, LeFloch, Launay, Letzelter, and Rostas (1987) | CO energy levels |
| FBLLT83 | Field, Benoist d'Azy, Lavollée, Lopez-Delgado, and Trainer (1983) | CO lifetimes |
| GCJ91 | Gilbert, Child, and Johns (1991) | $H_2O$ wavenumbers |
| H61a | Herzberg (1961a) | $CH_2$ wavelengths |
| HLM69 | Herzberg, Lagerqvist, and Malmberg (1969) | $C_2$ wavelengths |
| J63 | Johns (1963) | $H_2O$ wavelengths |
| K76 | Kurucz (1976) | CO wavelengths |
| KC89 | Kirby and Cooper (1989) | CO theoretical *f*-values |
| L84 | Lien (1984) | CH wavelengths, *f*-values |
| LvDWD82 | Langhoff, van Dishoeck, Wetmore, and Dalgarno (1982) | OH theoretical *f*-values |
| MD76 | Morton and Dinerstein (1976) | $H_2$ theoretical *f*-values |
| PRSR83 | Pouilly, Robbe, Schamps, and Roueff (1983) | $C_2$ theoretical *f*-values |
| S76 | Snow (1976) | summary of several molecules |
| SA73 | Sutherland and Anderson (1973) | OH lifetimes and *f*-values |
| SBT69 | Simmons, Bass, and Tilford (1969) | CO wavenumbers |
| SYBP80 | Smith, Yoshino, Black, and Parkinson (1980) | HCl experimental *f*-values |
| SYGB81 | Smith, Yoshino, Griesinger, and Black (1981) | $H_2O$ experimental *f*-values |
| TS72 | Tilford and Simmons (1972) | CO wavelengths |
| TG71 | Tilford and Ginter (1971) | HCl wavenumbers |
| TGV70 | Tilford, Ginter, and Vanderslice (1970) | HCl wavenumbers |
| vDB86 | van Dishoeck and Black (1986) | summary of several molecules |
| vDLD83 | van Dishoeck, Langhoff, and Dalgarno (1983) | OH theoretical *f*-values |
| vDvHD82 | van Dishoeck, van Hemert, and Dalgarno (1982) | HCl theoretical *f*-values |
| WH80 | Wang and Huang (1980) | OH experimental *f*-values |
| W68 | Wilkinson (1968) | HD wavenumbers |

# Updates

Cardelli, Savage and Ebbetts (1991) have measured interstellar absorption lines of Ga II, Ge II and Kr I in the spectrum of ζ Oph and Ge II in ξ Per. None of these species had been reported in *Copernicus* spectra, though two unidentified lines at 1164.275 and 1237.089 Å towards ζ Pup (Morton 1978) match wavelengths for Ge II. However, their strengths imply an abundance relative to hydrogen 23 times the solar system value.

There are some recent contributions to atomic data that deserve mention here. O'Brian and Lawler (1992) have measured accurate lifetimes for many levels of Si I. When used with the branching ratios of Smith *et al* (1987), the resultant *f*-values for UV multiplets 1, 2 and 7 are significantly more accurate than those adopted by Morton (1991) directly from Smith *et al.* These improved values will be useful if Si I is detected as interstellar absorption.

Wiese (1991 private communication) and his colleagues at NIST have compared the C I transition probabilities which Luo and Pradhan (1989) calculated for the opacity project (Mendoza, this symposum) with recent experimental measures and found reasonable agreement. Although the calculations are available for only the LS transitions, the quoted *f*-values for these should be better than the estimates from interstellar line strengths adopted by Morton (1991). Also, in that compilation the quoted values for multiplet 6 of C I mistakenly were taken from the theoretical calculations of Nussbaumer and Storey (1984). The preferred experimental values from Goldbach, Martin and Nollez (1989) are $A_{55}(1279.498) = 8.0 \times 10^5$, $A_{75}(1279.229) = 1.1 \times 10^7$ and $A_{53}(1279.056) = 1.8 \times 10^6\ s^{-1}$.

The latest calculations of strong lines by Kurucz (1989) also show good agreement with experiments, such as those for Fe I (O'Brian *et al* 1991), but the earlier results for C I published by Kurucz & Peytremann (1975) can have significant deviations from experimental values (Goldbach *et al* 1989).

New CO absorption cross sections measured by Eidelsberg *et. al* (1992) for the $A\ ^1\Pi$ - $X\ ^1\Sigma^+$ band give vibrational *f*-values consistent with those in Table 1 and extend the range to $v' = 12$. Also, Eidelsberg *et al* (1991) have published a useful atlas of the CO spectrum from 912 to 1152 Å, though Smith *et al* (1992) have demonstrated a systematic error of about 1 $cm^{-1}$ in the wavenumbers (10mÅ in the wavelengths).

Unfortunately, since the presentation of this paper in Buenos Aires, we have learned of the difficulties with the power distribution on one side of the GHRS. Power is no longer reaching the Echelle A detector with the LiF window, which extended sensitivity to $\lambda < 1150$ Å. Also the transfer of data from the second detector, with a $MgF_2$ window, is intermittent. Observing with the GHRS has been suspended until the problem with the $MgF_2$ data is understood. Even then the highest resolution of 3.2 km $s^{-1}$ will be available only longward of 1700 Å with Echelle B. From 1700 to 1150 Å a resolution of $2 \times 10^4$ or 15 km $s^{-1}$ will be obtainable with grating G160M, though there is hope that deconvolution algorithms will improve this to 7.5 km $s^{-1}$.

I wish to thank Dr. Daniel Durand and Dr. Stephen Morris of the Dominion Astrophysical Observatory for their assistance in obtaining the plots of GHRS spectra from the HST archive.

# References

Aashamar, K., and Luke, T.M. 1990, J. Phys., B23, L733.

Allison, A.C., and Dalgarno, A. 1970, Atomic Data, 1, 289.

Anders, E., and Grevesse, N. 1989, Geochim. Cosmochim. Acta, 53, 197.

Ansbacher, W., Li, Y., and Pinnington, E.H. 1989, Phys. Lett. A, 139, 165.

Biémont, E., Froese Fischer, C., Godefroid, M., Vaeck, N., and Hibbert, A. 1990, *Proc. Atomic Data and Oscillator Strengths for Astrophysics and Fusion Research* (Amsterdam, August 28-31, 1989), ed. J.E. Hansen (Amsterdam: North-Holland), 59.

Biémont, E., Hibbert, A., Godefroid, M. and Vaeck, N., and Fawcett, B.C. 1991a, ApJ, 375, 818.

Biémont, E., Baudoux, M., Kurucz, R.L., Ansbacher, W., and Pinnington, E.H. 1991b, A & A, 249, 539.

Boggess, A. *et al.* 1978, Nature, 275, 372.

Cardelli, J.A., Savage, B.D., Bruhweiler F.C., Smith, A.M., Ebbets, D.C., Sembach K.R., and Sofia, U.J. 1991, ApJ, 377, L57.

Cardelli, J.A., Savage, B.D., and Ebbets D.C. 1991, ApJ. Lett., (submitted).

Coxon, J.A. 1980, Can. J. Phys., 58, 933.

Curtis, L.J., Engman, B., and Martinson, I. 1976, Phys. Scripta, 13, 109.

Dabrowski, I. 1984, Can. J. Phys., 62, 1639.

Douglas, A.E. 1974, Can. J. Phys., 52, 318.

Eidelsberg, M., Benayoun, J.J., Viala, Y., and Rostas, F. 1991, A&A Suppl., 90, 231.

Eidelsberg, M., Roncin, J.-Y., Le Floch, A., Launay, F., Letzelter, C., and Rostas, J. 1987, J. Molec. Spectr., 121, 309.

Eidelsberg, M., 1992, J. Chem. Phys., (submitted).

Field, R.W., Benoist d'Azy, O., Lavolée, M., Lopez-Delgado, R., and Trainer, A. 1983, J. Chem. Phys., 78, 2838.

Ford, A.L. 1975, J. Molec. Spectr., 56, 251.

Gaupp, A., Kuske, P., and Andrä, H.J. 1982, Phys. Rev. A, 26, 3351.

Gawlik, W., Kowalski, J., Neumann, R., Wiegemann, H.B., and Winkler, K. 1979, J. Phys. B, 12, 3873.

Gilbert, R.D., Child, M.S., and Johns, J.W.C. 1991, Molec. Phys., 74, 473.

Goldbach, C., Martin, M., and Nollez, G. 1989, A&A, 221, 155.

Gosselin, R.N., Pinnington, E.H., and Ansbacher, W. 1988a, Nucl. Instr. Methods Phys. Res., B31, 305.

Gosselin, R.N., Pinnington, E.H., and Ansbacher, W. 1988b, Phy.Rev. A, 38, 4887.

Grevesse, N., Lambert, D.L., Sauval, A.J., van Dishoeck, E.F., Farmer, C.B., and Norton, R.H. 1990, A&A, 232, 225.

Herzberg, G. 1961a, Proc. Roy. Soc., A262, 291.

Herzberg, G. 1961b, Can. J. Phys., 39, 1511.

Herzberg, G., Lagerqvist, A., and Malmberg, C. 1969, Can. J. Phys., 47, 2735.

Hsu, D.K., and Hayden Smith, W., 1977, Spectroscopy Letters, 10, 181.

Jenkins, E.B., Drake, J.F., Morton, D.C., Rogerson, J.B., Spitzer, L., and York, D.G. 1973, ApJ, 181, L122.

Jenkins, E.B., Lees, J.A., van Dishoeck, E.F., and Wilcots, E.M. 1989, ApJ, 344, 785.

Johns, J.W.C. 1963, Can. J. Phys., 41, 209.

Johnson, B.C., Smith, P.L., and Knight, R.D. 1984, ApJ, 281, 477.

Kaufman, V., and Edlén, B. 1974, J. Phys. Chem. Ref. Data, 3, 825.

Keenan, F.P., Bates, B., Dufton, P.L., Holmgren, D.E., and Gilheany, S. 1990, ApJ, 348, 322.

Kirby, K., and Cooper, D. L., 1989, J. Chem. Phys., 90, 4895.

Knight, R.D. 1982, Phys. Rev. Lett., 48, 12.

Kono, A., and Hattori, S. 1982, J. Opt. Soc. Am., 72, 601.

Kurucz, R.L. 1976, Smithsonian Ap Obs. Special Rep., 374.

Kurucz, R.L. 1989, computer tapes.

Kurucz, R.L., and Peytremann, E. 1975, Smithsonian Astrophysical Observatory Special Report, 362.

Kwong, H.S., Johnson, B.C., Smith, P.L., and Parkinson, W.H. 1983, Phys. Rev. A, 27, 3040.

Kwong, H.S., Smith, P.L., and Parkinson, W.H. 1982, Phys. Rev. A, 25, 2629.

Langhoff, S.R., van Dishoeck, E.F., Wetmore, R., and Dalgarno, A. 1982, J. Chem. Phys., 77, (3), 1379.

Lee, L.C., and Guest, J.A. 1981, J. Phys. B, 14, 3415.

Letzelter, C., Eidelsberg, M., Rostas, F., Breton, J., and Thieblemont, B. 1987, Chem. Phys., 114, 273.

Lien, D.J. 1984, ApJ, 284, 578.

Luo, D., and Pradhan, A.K., 1989, J. Phy. B, 22, 3377.

Moore, C.E., 1950, NBS Circular 488.

Morton, D.C. 1975, ApJ, 197, 85.

Morton, D.C. 1978, ApJ, 222, 863.

Morton, D.C. 1991, ApJS, 77, 119.

Morton, D.C., and Dinerstein, H.L. 1976, ApJ, 204, 1.

Morton, D.C., and Spitzer, L. 1966, ApJ, 144, 1.

Nussbaumer, H., Pettini, M., and Storey, P.J. 1981, A&A, 26, 351.

Nussbaumer, H., and Storey, P.J. 1984, A&A, 140, 383.

O'Brian, T.R., and Lawler, J.E. 1992, Phys. Rev. A, (submitted).

O'Brian, T.R., Wickliffe, M.E., Lawler, J.E., Whaling, W., and Brault, J.W. 1991, J. Opt. Soc. Am., B8, 1185.

Pouilly, B., Robbe, J.M., Schamps, J., and Roueff, E. 1983, J. Phys. B, 16, 437.

Rogerson, J.B., Spitzer, L., Drake, J.F., Dressler, K., Jenkins, E.B., Morton, D.C., and York, D.G. 1973, ApJ, 181, L97.

Rottman, G.J., Hassler, D.M., Jones, M.D., and Orrall, F.Q. 1990, ApJ, 358, 693.

Savage, B.D., Cardelli, J.A., Bruhweiler, F.C., Smith, A.M., Ebbets, D.C., and Sembach, K.R. 1991, ApJ, 377, L53.

Simmons, J.D., Bass, A.M., and Tilford, S.G. 1969, ApJ, 155, 345.

Smith, A.M., Bruhweiler, F.C., Lambert, D.L., Savage, B.D., Cardelli, J.A., Ebbets, D.C., Lyu, C-H., and Sheffer, Y. 1991, ApJ, 377, L64.

Smith, P.L., Huber, M.C.E., Tozzi, G.P., Griesinger, H.E., Cardon, B.L., and Lombardi, G.G. 1987, ApJ, 322, 573.

Smith, P.L., Kwong, H.S., Johnson, B.C., and Parkinson, W.H. 1983, Bul. Am. Astron. Soc., 15, 703.

Smith, P.L., Johnson, B.C., Kwong, H.S., and Parkinson, W.H. 1984, Phys. Scripta, 48, 88.

Smith, P.L., Yoshino, K., Black, J.H., and Parkinson, W.H. 1980, ApJ, 238, 874.

Smith, P.L., Yoshino, K., Griesinger, H.E., and Black, J.H. 1981, ApJ, 250, 166.

Smith, P.L., Yoshino, K., Stark, G., and Shettle, A. 1992, A & A, (submitted).

Snow, T.P. 1975, ApJ, 201, L21.

Snow, T.P. 1976, ApJ, 204, L127.

Snow, T.P. 1980, IAU Symposium, 87, p. 247.

Stone, M.E., and Morton, D.C. 1967, ApJ, 149, 29.

Sutherland, R.A., and Andersen, R.A. 1973, J. Chem. Phys., 58, 1226.

Theodosiou, C.E. 1986, J. Opt. Soc. Am. B, 3, 1107.

Theodosiou, C.E. 1989, Phys. Rev. A, 39, 4880.

Theodosiou, C.E., and Curtis, L.J. 1988, Phys. Rev. A, 38, 4435.

Tilford, S.G., and Ginter, M.L. 1971, J. Molec. Spectr., 40, 568.

Tilford, S.G., Ginter, M.L., and Vanderslice, J.T. 1970, J. Molec. Spectr., 33, 505.

Tilford, S.G., and Simmons, J.D. 1972, J. Phys. Chem. Ref. Data, 1, 147.

van Dishoeck, E.F., and Black, J.H. 1986, ApJS, 62, 109.

van Dishoeck, E.F., Langhoff, S.R., and Dalgarno, A. 1983, J Chem. Phys., 78, 4552.

van Dishoeck, E.F., van Hemert, M.C., and Dalgarno, A. 1982, J. Chem. Phys., 77, 3693.

Wang, C.C., and Huang, C.M. 1980, Phys. Rev. A, 21, 1235.

Whiting, E.E., Schadee, A., Tatum, J.B., Hougen, J.T., and Nicholls, R.W. 1980, J. Mol. Spectr., 80, 249.

Wiese, W.L. 1987, Phys. Scripta, 35, 846.

Wilkinson, P.G. 1968, Can. J. Phys., 46, 1225.

# 6

# Atomic and Molecular Data Needed for Analysis of Infrared Spectra from ISO and SIRTF

P.F. Bernath[1]

ABSTRACT The satellites ISO and SIRTF are cryogenically-cooled infrared observatories. These telescopes will provide a tremendous increase in sensitivity and will, therefore, require new laboratory data to support their missions. A survey of some of the necessary atomic and molecular data is presented.

## Introduction

The Infrared Space Observatory (ISO) and the Space Infrared Telescope Facility (SIRTF) are cryogenically-cooled infrared observatories which will operate in Earth orbit. They are designed to operate in the 2-200 micron spectral region with moderate resolving power (less than 10,000 for ISO, less than 2,000 for SIRTF) for spectroscopy, photometry and imaging. Because the telescopes are cryogenically-cooled and above the Earth's atmosphere, they will provide a spectacular increase in sensitivity (up to three orders of magnitude) unimpeded by telluric absorption.

The performance of current infrared telescopes is limited in wavelength by strong absorption in the Earth's atmosphere. Indeed, for the far infrared region from 40-300 microns (33-250 $cm^{-1}$) no radiation penetrates the Earth's atmosphere. To shorter wavelengths only selected windows between the strong $CO_2$ and $H_2O$ absorption bands are available.

Even more deleterious to ground-based infrared observations than the limited wavelength coverage is the tremendous background infrared flux. The bulk of this unwanted radiation comes from the telescope itself and from the Earth's atmosphere. Modern ground based infrared astronomy has become so sophisticated that signals from objects $10^{-6}$ times weaker than the background at long wavelengths (10 microns) can be routinely detected. However, shot-noise from the background radiation is unavoidable, so the ultimate sensitivity can be improved only by eliminating the background. Cooling the telescope to less than 4K and placing the telescope above the Earth's atmosphere decreases the infrared

[1] Departments of Chemistry and Physics, University of Waterloo, Waterloo, Ontario, Canada N2L 3G1

background radiation by seven orders of magnitude (Rieke et al. 1986).

The infrared region provides a unique window to view the cosmos. In general, infrared radiation is associated with relatively cool matter with temperatures from 10-1000K. All molecules have infrared transitions, including homonuclear diatomics such as $H_2$ which have weak quadrupole vibration-rotation transitions (Scoville et al. 1982). Hot solids also emit radiation strongly in the infrared. Molecules and solids are key components in regions of star formation, planetary atmospheres and circumstellar shells.

The most important property of infrared radiation is the ability to penetrate clouds of cosmic dust. Dust particles obscure some of the most important regions in the Universe, including the centres of galaxies, star-forming regions and circumstellar shells. The strong inverse dependence of the scattering cross-section on wavelength allows infrared radiation to escape from dust enshrouded objects, but traps visible and ultraviolet radiation. Infrared radiation is thus a unique probe of "hidden" matter in the Universe (Rieke et al. 1986).

ISO and SIRTF will also provide excellent spectra of objects in our own solar system. Most of the parent molecules in comets (eg. $H_2O$, $CH_4$ and $NH_3$) have infrared spectra (Hughes 1982, Mumma et al. 1986). The spectra of the planets, particularly in the far infrared, can be monitored over extended periods of time (Larson 1980).

Both the ISO and SIRTF can be viewed as successors to the very successful Infrared Astronomy Satellite (IRAS 1983). IRAS was designed for a quick survey of the entire sky in four infrared bands, while ISO and SIRTF are true infrared observatories.

# ISO

The acronym ISO stands for Infrared Space Observatory. ISO is a large satellite (5.3 meters by 2.3 meters in diameter, 2400 kg in mass at launch) built by the European Space Agency (ESA) for launch by Ariane rocket in 1993 (Kessler 1991, Metcalfe and Kessler, 1989). The satellite will be in Earth orbit with a 24 hour period with a 1000 km perigee and 70,000 km apogee. Operations will be directed from ESA's Villafranca ground station near Madrid, Spain. The satellite will be in contact with the ground station for 14 hours per day and it is hoped that a partner can be found to build another ground station to complete the coverage. The satellite is, in fact, a large cryostat with the telescope and instrumentation inside. The telescope and instrumentation are cooled to below 4K by 2300 litres of superfluid helium. The expected lifetime of the satellite is 18 months, limited by the supply of cryogen.

The telescope is a 60 cm, f/15, Ritchey-Chrétien design with gold-coated fused silica primary and secondary mirrors. The telescope will provide diffraction limited performance at 5 microns and the pointing accuracy is 2.7 arc seconds. The design has been carefully optimized with a sunshade and light baffles to minimize stray infrared light.

There are four scientific instruments mounted at the Cassegrain focus, each one occupying an 80° segment of the total field of view. The 20 arc minute field of view is divided into four equal sectors with a pyramidal mirror. Each instrument simultaneously receives a 3 arc minute field of a different part of the sky. In order to view the same target with different instruments, the telescope has to be repointed. The four scientific instruments on ISO are: a camera, an imaging photopolarimeter, a long wavelength spectrometer, and a short wavelength spectrometer.

The camera, ISOCAM, uses two small format (32x32) infrared arrays made of InSb (2.5 to 5.5 microns) and Si:Ga (4 to 17 microns) for imaging. The field of view is up to 3 minutes wide, with a variety of filters available to restrict the spectral bandwidth.

The imaging photopolarimeter, ISOPHOT, is a complex and flexible instrument with a variety of single detectors, small mosaic detector arrays, filters and polarizers. Photometry (absolute photometric accuracy < 5%), infrared imaging and polarization measurements are possible in the 2.5 to 200 micron range.

The long wavelength spectrometer, LWS, covers the 45-180 micron (55-222 $cm^{-1}$) spectral range with a resolving power of ~100-200. The instrument is a relatively standard scanning grating spectrometer. Detector arrays are not used, but the 45-180 micron spectral range at the output plane is divided into ten parts with a Ge:Ga detector for each channel. The ten detectors are in a line and simultaneously detect the output of the spectrometer to obtain a multiplex advantage.

While a maximum resolving power $\Delta\lambda/\lambda$ of 200 is suitable for survey work and solid state spectroscopy at long wavelengths, higher resolution is very helpful for atomic and molecular work. The resolving power can, therefore, be increased to 10,000 by insertion of one of two Fabry-Perot étalons (one for the 45-90 micron range, the second for 90-180 microns). When inserted in the optical path, the spacing between the two mirrors of the étalon is changed under servo control to scan the wavelength in high resolution mode.

The short wavelength spectrometer, SWS, covers the 2.5 to 45 micron (222-4000 $cm^{-1}$) spectral range with a resolving power of about 1000. The SWS consists essentially of two parallel scanning spectrometers, one covering the 2.4-13 micron range and the other the 12-45 micron part of the spectrum. In order to gain an additional multiplex advantage the output of the spectrometers falls on 48 individual detectors in the exit plane: 12 InSb for 2.4-4.1 microns, 12 Si:Ga for 4.1-13 microns, 12 Si:P for 11-29 microns and 12 Ge:Be for 27-45 microns. Similar to the operation of the LWS, the output of the 48 individual detectors is monitored as the spectrometers are scanned, with each detector devoted to a different section of the spectrum in the 2.5-45 micron range.

The large format infrared arrays which are just becoming available will be used for SIRTF (but not ISO). They will greatly improve the sensitivity. Large format two dimensional arrays will eliminate scanning and allow the instrument to operate as a spectrograph.

The SWS of ISO also has a high resolution mode using Fabry-Perot étalons. Two étalons (one for each spectrometer of SWS) can be selected for a resolving power of about 20,000. A separate optical path and detectors are provided for the high resolution mode. Like the LWS, the étalons are operated by scanning the separation between the two mirrors of the étalon under servo control.

## SIRTF

SIRTF is an acronym for Space Infrared Telescope Facility and is the infrared component of NASA's Great Observatories Program (Rieke et al. 1986, Werner 1990). The Great Observatories Program for astrophysics research consists of four large satellites each covering a different portion of the electromagnetic spectrum. The Gamma Ray Observatory (GRO), the Advanced X-Ray Astrophysics Facility (AXAF), the Hubble Space Telescope (HST) and the Space Infrared Telescope Facility (SIRTF) will offer almost complete spectral coverage from the far infrared to the gamma ray part of the electromagnetic spectrum.

The SIRTF project is at a much earlier stage of development that ISO. SIRTF is scheduled to be a new start in 1993 for a launch in the late 1990's. The satellite will be launched by a Titan IV/Centaur rocket in a high, circular orbit of 100,000 kilometers. This distant orbit places the satellite above the Van Allen radiation belts and minimizes the effect of the Earth on the observatory. The small angular diameter (less than 7 degrees) of the Earth as seen from the satellite minimizes the thermal load on the satellite and permits long, continuous observations of a single target for 50 hours or more. The on-target efficiency will exceed 80%. The 26 meter dishes of the Deep Space Network on Earth will receive the data transmitted by SIRTF.

Like ISO, SIRTF is a cryogenic telescope cooled to less than 4K by superfluid helium. The expected lifetime of SIRTF is five years. The primary mirror of the telescope will have a diameter of 95 centimeters. The telescope will provide diffraction limited performance at 3 microns (and longer) and the pointing accuracy is 0.25 arcseconds. There are three scientific instruments on SIRTF: an Infrared Array Camera (IRAC), a Multiband Imaging Photometer (MIPS) and an Infrared Spectrograph (IRS).

The Infrared Array Camera, IRAC, will provide wide field imaging using large format infrared array detectors in the 2-30 micron (333-5000 $cm^{-1}$) region. Filters will be available to limit the spectral bandpass. The current plans are to use three array detectors: a 256x256 pixel InSb (1.8-5.3 microns) and two 128x128 pixel Si:As arrays (5.3-14 microns and 14-30 microns).

The Multiband Imaging Photometer for SIRTF, MIPS, will provide imaging and photometry in the spectral range 15-700 microns (14-666 $cm^{-1}$). The long wavelength limit of MIPS may be extended to take advantage of the very low far infrared background radiation in high Earth orbit. Again photoconductive detector arrays will be used but for long wavelengths ($\lambda$ > 200 microns) bolometers must be used. A special adiabatic demagnetization refrigerator will be used to cool the bolometers to 0.1K for maximum far infrared sensitivity.

The Infrared Spectrograph, IRS, will cover the 2.5-200 micron (50-4000 $cm^{-1}$) range with several spectrographs. In this range there will be a low resolution, high-sensitivity mode with a resolving power ($\lambda/\Delta\lambda$) of about 100 and a higher resolution mode with a resolving power of about 2000. The "cross-dispersion" design of the spectrograph will allow large format detector arrays such as the 128x128 element Si:As detector (4-30 microns) to be used.

In some sense ISO is the follow-up to IRAS, and SIRTF will follow ISO. This gives SIRTF a tremendous advantage in terms of sensitivity. Rapid progress is anticipated in the development of large format infrared arrays in the next few years. The goal of the SIRTF project is to reach natural background limited performance. The background radiation in the infrared is from zodiacal scattering and emission at shorter wavelengths and galactic emission at longer wavelengths (Rieke et al. 1986). At very long wavelengths the 3K cosmic background and emission from the spacecraft also contribute.

One advantage of ISO over SIRTF (as currently planned) is the possibility of higher spectral resolving power (20,000 versus 2,000). The general rule in spectroscopy is that the resolution of the spectrometer should be increased until the linewidth of the atomic or molecular species is reached. Unfortunately there is a trade-off between sensitivity and resolving power so that it is possible to set the resolution too high and observe nothing!

For gas-phase atomic and molecular features the maximum resolution of 2,000 proposed for SIRTF may not be high enough. If the design sensitivity for the spectrograph is achieved then the "confusion limit" may be reached. For example in millimeter wave astronomy the sensitive receivers currently available result in the detection of a confusing forest of lines, mostly due to interstellar "pollutants" such as methyl formate. Only very high resolution spectra and the availability of accurate laboratory frequencies allow progress to be made. A similar problem may occur for SIRTF (and ISO) where higher resolution would be helpful in sorting out a jumble of overlapping lines. Direct laboratory measurements of the infrared spectra of the atoms and molecules will prove indispensable in simulating the spectra observed by SIRTF and ISO.

## Atomic Spectroscopy

Atomic transitions are prominent in the infrared spectra of relatively hot objects such H I regions, diffuse clouds and planetary nebulae. In cooler, denser regions the atoms tend to be associated in the form of molecules or on grains. Atoms are found in some parts of dark molecular clouds where molecules are dissociated into atoms by ultraviolet radiation or shock waves.

The recombination lines of hydrogen in the infrared are well known. Infrared transitions of less abundant elements such as Fe, Ne, O, etc. should be observed with the increase in sensitivity available with ISO and SIRTF. Laboratory data on infrared spectra of neutral atoms and ions are very sparse. Recently, spectra of a few elements such as Fe (Biémont et al. 1985) and Mg (Biémont and Brault 1986) have been extended from the visible to 5 microns (1800 $cm^{-1}$) by Fourier transform spectroscopy. Atomic spectra recorded with the McMath Fourier transform spectrometer of the National Solar Observatory are notable. In general, infrared atomic spectra of most elements are unknown, as are the corresponding oscillator strengths of the lines. Fortunately, most of the lines can be predicted from the known atomic energy levels and the oscillator strengths estimated by ab inito calculations. Direct observation of the infrared atomic transitions is, however, desirable since the energy levels of most elements are far from complete.

The problems associated with inadequate atomic data in the infrared region is illustrated by the discovery of mysterious solar emission lines at 12 microns (Brault and Noyes 1983, Murcray et al. 1981). The identification of the strong, magnetically sensitive lines at 811.575 $cm^{-1}$ and 818.058 $cm^{-1}$ as well as numerous weaker lines was difficult because no laboratory data was available. Chang and Noyes (1983) were able to identify the transitions as high-$l$ Rydberg transitions of neutral magnesium and aluminum. Finally, in 1990 direct laboratory observations of the magnesium transitions were made by Lemoine et al. (1990) by diode laser spectroscopy, confirming the assignments.

Extensive infrared observations of atomic spectra by laser and Fourier transform spectroscopy will be necessary to interpret the ISO and SIRTF data. The acquisition of the necessary data is slowed both by the lack of funding and by the small number of active atomic spectroscopists.

In addition to the strong allowed infrared transitions, a number of important forbidden transitions have already been detected (Watson 1984). The most astronomically significant transitions are the magnetic dipole transitions between fine structure components in the far infrared region. The [C I] $^3P_1$-$^3P_0$, [C II] $^2P_{3/2}$-$^2P_{1/2}$ and [O I] $^3P_1$-$^3P_2$, $^3P_0$-$^3P_1$ transitions at 609.133, 157.73, 63.1837 and 145.526 microns, respectively, have been detected with NASA's Kuiper Airborne Observatory. These transitions are important probes of cloud density and temperature as well as acting as the main coolants in diffuse clouds.

Excellent laboratory data for the [C I] transition are available (Yamamoto and Saito 1991) but similar data for less abundant atoms in all stages of ionization are needed. Transitions in S, Si (Inguscio et al. 1984) and N (Cooksy et al. 1986) will certainly be found (and many additional atoms are possible) by ISO and SIRTF. Although intramultiplet transitions can be predicted with modest accuracy from existing optical and ultraviolet transitions, direct precise measurements are needed.

# Molecular Spectroscopy

## Electronic Transitions

Surprisingly many molecules have electronic transitions at infrared wavelengths. Molecular ions can recombine with electrons to make highly excited Rydberg states which can emit a cascade of infrared radiation as they cool. These transitions are the molecular analogs of hydrogen recombination lines. In the laboratory these infrared transitions have been directly observed only for a few molecules such as ArH, KrH and XeH (Herzberg 1987, Douay et al. 1988a) but they exist, in principle, for all molecules (although the states may be predissociated). ISO and SIRTF may discover infrared Rydberg transitions of molecules in sources such as H I regions and planetary nebulae. Laboratory spectroscopy of Rydberg transitions of molecules is at a very early stage and much experimental (and theoretical) work on molecules such as CH, NH and OH is necessary.

Many free radicals, ions and transient molecules also have infrared electronic transitions. Perhaps the best known astronomical examples are CN and $C_2$. Although the red system ($A^2\Pi$ - $X^2\Sigma^+$) of CN and the Phillips system ($A^1\Pi_u$ - $X^1\Sigma^+{}_g$) of $C_2$ have been detected in many astronomical sources including comets, interstellar clouds, the sun and stellar atmospheres, the laboratory spectroscopy remains surprisingly incomplete. For example, numerous CN bands are found in the 3-5 micron interval in stellar atmospheres (Hinkle 1991) but laboratory comparison spectra are not published. Similar infrared transitions exist for the isovalent CP (Ram and Bernath 1987) and SiN (Yamada et al. 1988) molecules.

For $C_2$ two new strong infrared electronic transitions ($B'\ {}^1\Sigma^+{}_g$ - $A^1\Pi_u$ and $B^1\Delta_g$ - $A^1\Pi_u$) were found (Douay et al. 1988b) in the 3-5 micron range. These new $C_2$ transitions should be found in various sources including comets and stellar atmospheres. Many other molecules including $C_2H$ (Curl et al. 1985), $C_4H$, $C_6H$ and $C_5$ are expected to have infrared electronic transitions for which laboratory data is incomplete ($C_2H$) or unavailable. A new triplet transition of $C_3$ was recently identified near 6500 $cm^{-1}$ (Sasada et al. 1991).

SiC is an elusive molecule isovalent with $C_2$. Recently infrared electronic transitions (Figure 1) of SiC were found in the laboratory (Bernath et al. 1988, Brazier et al. 1989). SiC was also found in the prototypical obscured carbon star, IRC + 10216, by radio observations and in the laboratory by millimeter wave spectroscopy (Cernicharo et al. 1989).

As illustrated by the recent work on $C_2$ and SiC, infrared electronic spectroscopy of simple molecules containing the elements H, C, N, O, Si, P and S is very incomplete. Advances in infrared laser and Fourier transform spectroscopy allow infrared electronic transitions to be observed in the laboratory. Since electronic transition dipole moments are typically 1 Debye, and vibration-rotation transition dipole moments are typically 0.1D, infrared electronic transitions are much stronger (~100 times) than vibration-rotation

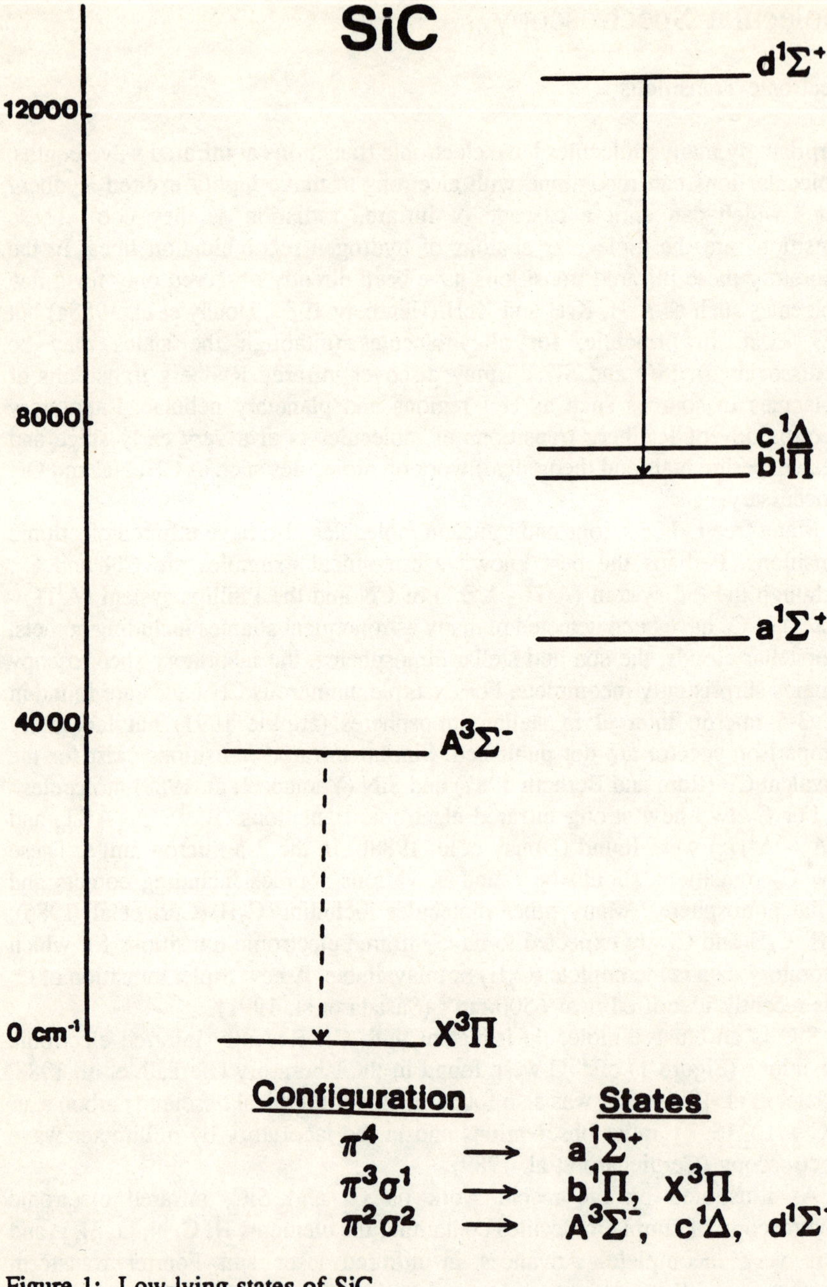


Figure 1: Low-lying states of SiC.

transitions if all else is equal. This means that ISO and SIRTF may detect infrared electronic transitions in sources such as IRC + 10216 where transient molecules are found.

Vibration-Rotation Transitions

All molecules have vibration-rotation transitions so that infrared radiation is a unique probe for establishing chemical abundances and distributions. Even homonuclear diatomic molecules such as $H_2$, which have no electric dipole allowed transitions, can be detected in energetic regions by quadrupole vibration-rotation transitions.

For the "standard" molecular species, such as $CO_2$, $CH_4$, $H_2O$ and $NH_3$, there are excellent laboratory spectra recorded at room temperature. However, the temperature of molecules in various astronomical sources can range from 10K in dark clouds to 3000K in the atmospheres of cool stars. The spectra of low temperature molecules present no problems since they can be calculated easily from the room temperature laboratory data. The high temperature spectra present more of a problem since the spectroscopic constants used to represent the room temperature data do not reliably predict the high J transitions required to simulate a spectrum at 3000K. The problem is particularly acute for light hydrides such as $H_2O$. For example, there are a large number of unassigned absorption features in Mira variable stars which could be due to hot $H_2O$ but no laboratory data is available (Hinkle 1991). This will, undoubtedly, be a problem for the high quality data produced by ISO and SIRTF. Spectra of hot stable molecules should be recorded in the laboratory to help simulate spectra of abundant species such as HCN, HCCH, $CH_4$, $NH_3$ and $H_2O$.

At least some laboratory data are available for stable chemical species such as $CH_4$, but for transient species, spectroscopic data are very sparse. The success of millimeter wave astronomy in identifying new molecules serves as a guide. A large number of transient species such as long chain cyanoacetylenes, $C_3H_2$, $HCO^+$, $C_2S$, $H_2CCC$, etc. (Herbst 1990) have been detected in many sources including molecular clouds and circumstellar envelopes. The high spectral resolution of the astronomical observations proved very helpful in making assignments. Many of these transient molecules were discovered in space before they were confirmed by laboratory millimeter wave spectroscopy.

High sensitivity infrared observations of many sources by ISO and SIRTF will be like looking into a "chemical soup". Laboratory infrared observations of the numerous species present will be vital in simulating the observed spectra. The moderate spectral resolution available in ISO and, particularly, SIRTF makes laboratory data essential. At very high resolution the infrared line positions (like microwave line positions) give a unique spectral signature to a molecule. At moderate resolution the lines from many species tend to be jumbled and overlapped, making unique identifications more difficult. The Voyager infrared data from the outer planets are a good example. With good laboratory data, convincing simulations were possible, and lead to the identification of $C_4H_2$,

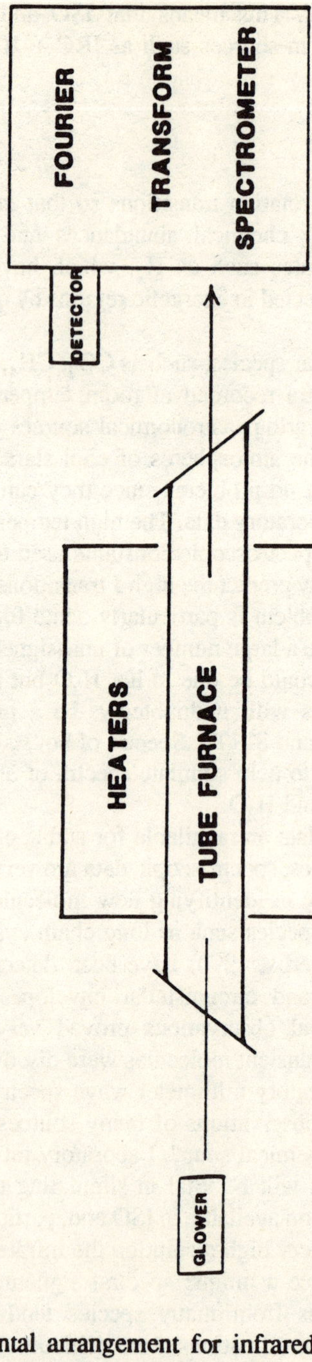


**Figure 2: Experimental arrangement for infrared Fourier transform emission spectroscopy of high temperature molecules.**

$HC_3N$ and $C_2N_2$ in Titan's atmosphere (Kunde et al. 1981).

The first priority should be the infrared detection in the laboratory of the transient molecules already detected by millimeter wave astronomy (Herbst 1990). Some infrared spectra of molecules such as $C_4H$ (Shen et al. 1990) isolated in argon matrices are available but the required gas-phase spectra are lacking. A little progress has been made recently, for example, with the laboratory infrared detection of $C_3H_2$ by Hirahara, Masuda and Kawaguchi (1991).

The infrared spectroscopy of molecular ions is, by contrast, in relatively good shape because of the extensive application of velocity modulation techniques (Gudeman and Saykally 1984). Infrared spectra of $NH_4^+$, $HCO^+$, $C_2H_2^+$ and $H_3^+$ (and many other ions) have been recorded in several groups. The pioneering efforts of the groups of Oka and Saykally are notable in this area. Recently $H_3^+$ has been detected in the atmosphere of Jupiter (Drossart et al. 1989) so that the detection of additional ions by ISO and SIRTF is anticipated.

The gas-phase infrared spectra of large molecules are also important. For example the unidentified infrared emission bands (Puget and Léger 1989), first identified in the planetary nebula NGC 7027, have been attributed to the emission from polycyclic aromatic hydrocarbons (PAHs). It has been pointed out that a mixture of PAHs is expected and that the PAHs in these sources should be ionized and dehydrogenated. ISO and SIRTF will provide greatly improved spectra of these features and will, undoubtedly, find similar new features. Laboratory spectra, particularly gas-phase spectra, are essential in confirming the PAH hypothesis.

Related to the PAH hypothesis is the possibility of finding large pure carbon molecules in space. The small pure carbon chain molecules $C_3$ (Hinkle et al. 1988) and $C_5$ (Bernath et al. 1989) were recently detected in the circumstellar shell, IRC + 10216. $C_3$ and $C_5$ were found by high resolution infrared absorption spectroscopy with the Fourier transform spectrometer at the 4 meter telescope operated by Kitt Peak National Observatory. The vibration-rotation spectra of $C_3$ and $C_5$ at 5 microns were strong and consistent with a temperature about 40K. The origin of these molecules is not clear. They could be the building blocks or the photofragments of carbonaceous materials.

The discovery of $C_3$ and $C_5$ makes the possibility of finding larger species $C_7$, $C_9$ or $C_{60}$ quite tantalizing. The spheroidal fullerenes such as $C_{60}$ may be present in astronomical sources. Infrared spectroscopy is vital in identifying pure carbon molecules because, like $CH_4$ and $CO_2$, they lack dipole moments, so the pure rotational spectra are very weak. The electronic spectra are not very diagnostic, even if they could be observed in the obscured objects where carbon-containing molecules are expected to be found.

The technique of infrared emission spectroscopy is very powerful for recording the high temperature spectra of both small and large molecules as well as solids. The technique is very simple (Figure 2): light from a hot cell is sent into the emission port of a Fourier transform spectrometer. Spectra can be recorded in absorption (glower on) and emission (glower off) under the same

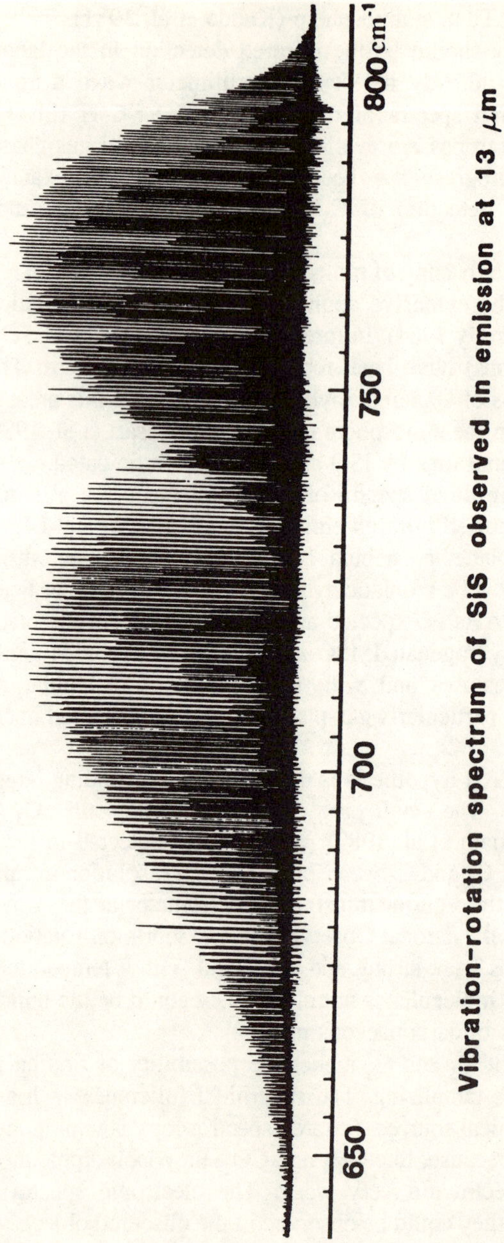


Figure 3: High resolution emission spectrum of SiS recorded at 1000°C with the McMath Fourier transform spectrometer of the National Solar Observatory.

experimental conditions. For example, the SiS molecule was observed at high resolution in emission at 13 microns (Figure 3) from the reaction of Si with $SiS_2$ (Frum et al. 1990). Some of the SiS vibration-rotational line positions matched those observed (Jennings et al. 1990) in the infrared spectrum of IRC + 10216. Although SiS had been observed in IRC + 10216 by microwave spectroscopy, our laboratory measurements confirmed the infrared detection.

Fourier transform emission spectroscopy can also be used to observe large molecules such as $C_{60}$ or PAH molecules. By heating up a sample of $C_{60}$ in our tube furnace to 700°C an infrared emission spectrum was recorded (Frum et al. 1991). Although $C_{60}$ has 174 normal modes of vibration, it has such high symmetry ($I_h$) that only four modes are infrared-active. The solid state infrared spectrum was known, but our measurements provided the gas-phase band positions for a search for $C_{60}$ in extraterrestrial sources.

Similar work needs to be carried out for many other small metal oxides, nitrides, carbides and hydrides which may be observed by ISO and SIRTF. Gas-phase spectra of large molecules such as PAHs, pure carbon molecules, hydrocarbon radicals and heterocyclic molecules also need to be recorded. As a result of work in the Saykally group infrared spectra of many of the small pure carbon molecules, $C_n$, are known (Heath et al. 1991), but much work remains particularly in the mid and far infrared regions.

## Pure Rotational Transitions

Pure rotational transitions of light molecules occur in the far infrared region. Since this region is completely obscured by atmospheric absorption, ISO and SIRTF will certainly make new discoveries. Pure rotational transitions of neutral molecules such as CO, HD, CH, NH, OH, HCl, $H_2O$ and $NH_3$ will be detected. The far infrared transitions of CO, OH and $NH_3$ have already been detected in Orion by measurements with the Kuiper Airborne Observatory (Watson 1984).

Molecular ions such as $CH^+$, $NH^+$, $HeH^+$, $OH^+$ and $H_2O^+$ may also be observed by ISO and SIRTF. The far infrared spectroscopy of these transient species in the laboratory is at a very early stage. Some measurements on OH (Blake et al. 1986), NH, CH, $OH^+$ (Bekooy et al. 1986) and $NH^+$ (Verhoeve et al. 1986) are available from far infrared laser measurements. Fortunately, infrared or optical measurements are available for molecules like $HeH^+$ but, as always, direct far infrared measurements are more accurate and should be carried out.

Van der Waals molecules may be found in the atmospheres of the outer planets. For example, the far infrared spectrum of $(H_2)_2$ has been tentatively identified in the Voyager spectra of Jupiter and Saturn (McKellar 1984). New laboratory far infrared spectra of $(H_2)_2$ and other van der Waals molecules are being recorded by McKellar at the National Research Council in Ottawa. ISO and SIRTF will provide excellent spectra of the dense atmospheres of the outer planets, so that the far infrared work on pressure-induced absorption and van der Waals systems is important.

## Acknowledgements

I thank the National Solar Observatory and Kitt Peak National Observatory for generous allocations of observing time to make some of the measurements described in this article. I also thank the NASA Origins of the Solar System Research Program for support. Acknowledgement is made to the Petroleum Research Fund, administered by the American Chemical Society for partial support of this work. I also thank G. Rieke, M. Kessler and M. Huber for information about SIRTF and ISO.

## References

Bekooy, J.P., Verhoeve, P., Meerts, W.L., and Dymanus, A. 1985, J. Chem. Phys., **82**, 3868.

Biémont, E., Brault, J.W., Delbouille, L., and Roland, G. 1985, Astron. Astrophys. Suppl., **61**, 107.

Biémont, E., and Brault, J.W. 1986, Phys. Scripta, **34**, 751.

Bernath, P.F., Rogers, S.A., O'Brien, L.C., Brazier, C.R., and McLean, A.D. 1988, Phys. Rev. Lett., **60**, 197.

Bernath, P.F., Hinkle, K.H., and Keady, J.J. 1989, Science, **244**, 562.

Blake, G.A., Farhoomand, J., and Pickett, H.M. 1986, J. Mol. Spectrosc., **115**, 226.

Brault, J., and Noyes, R. 1983, Ap. J., **269**, L61.

Brazier, C.R., O'Brien, L.C., and Bernath, P.F. 1989, J. Chem. Phys., **91**, 7384.

Chang, E.S., and Noyes, R.W. 1983, Ap. J., **275**, L11.

Cernicharo, J., Gottlieb, C.A., Guélin, M, Thaddeus, P., and Vrtilek, J.M. 1989, Ap. J., **341**, L25.

Cooksy, A.L., Hovde, D.C. and Saykally, R.J. 1986, J. Chem. Phys., **84**, 6101.

Curl, R.F., Carrick, P.G., and Merer, A.J. 1985, J. Chem. Phys., **82**, 3479.

Douay, M., Rogers, S.A., and Bernath, P.F. 1988a, Mol. Phys., **64**, 425.

Douay, M., Nietmann, R. and Bernath, P.F. 1988b, J. Mol. Spectrosc., **131**, 261.

Drossart et al. 1989, Nature, **340**, 539.

Frum, C.I., Engleman, R., and Bernath, P.F. 1990, J. Chem. Phys., **93**, 5457.

Frum, C.I., Engleman, R., Hedderich, H.G., Bernath, P.F., Lamb, L.D., and Huffman, D.R. 1991, Chem. Phys. Lett., **176**, 504.

Gudeman, C.S., and Saykally, R.J. 1984, Ann. Rev. Phys. Chem., **35**, 387.

Heath, J.R., Van Orden, A., Kuo, E., and Saykally, R.J. 1991, Chem. Phys. Lett., **182**, 17.

Herbst, E. 1990, Angew. Chemie, **25**, 595.

Herzberg, G. 1987, Ann. Rev. Phys. Chem., **38**, 27.

Hinkle, K.H., Keady, J.J., and Bernath, P.F. 1988, Science, **241**, 1319.

Hinkle, K.H. 1991, personal communication.

Hirahara, Y., Masuda, A., and Kawaguchi, K. 1991, J. Chem. Phys., **95**, 3975.

Hughes, D.W. 1982, Contemp. Phys., **23**, 257.

Inguscio, M., Evenson, K.M., Beltrán-López, V., and Ley-Koo, E. 1984, Ap. J., **278**, L127.

IRAS, The Infrared Astronomical Satellite, 1983, Nature, **303**, 287.

Jennings, D., Keady, J., Boyle, R.J., and Weidemann, G.R. 1990, private communication.

Kessler, M.F. 1991, Adv. Space Res., **11**, 277.

Kunde, V.G., Aikin, A.C., Hanel, R.A., Jennings, D.E., Maguire, W.C., and Samuelson, R.E. 1981, Nature, **292**, 686.

Larson, H.P. 1980, Ann. Rev. Astron. Astrophys., **18**, 43.

McKellar, A.R.W. 1984, Can. J. Phys., **62**, 760.

Lemoine, B., Petit, D., Destombes, J.L., and Chang, E.S. 1990, J. Phys. B., **23**, 2217.

Metcalfe, L., and Kessler, M.F. 1989, Scientific Capabilities of the ISO Payload, European Space Agency Report, SIO-SSD-8805.

Mumma, M.J., Weaver, H.A., Larson, H.P., Davis, D.S., and Williams, M. 1986, Science, **232**, 1523.

Murcray, F.J., Goldman, A., Murcray, F.H., Bradford, C.M., Murcray, D.G., Coffey, M.T., and Mankin, W.G. 1981, Ap. J., **247**, L97.

Puget, J.L., and Léger, A. 1989, Ann. Rev. Astron. Astrophys., **27**, 161.

Ram, R.S., and Bernath, P.F. 1987, J. Mol. Spectrosc., **122**, 282.

Rieke, G.H., Werner, M.W., Thompson, R.I., Becklin, E.E., Hoffmann, W.F., Houck, J.R., Low, F.W., Stein, W.A., and Witteborn, F.C. 1986, Science, **231**, 807.

Sasada, H., Amano, T., Jarman, C., and Bernath, P.F. 1991, J. Chem. Phys., **94**, 2401.

Scoville, N.Z., Hall, D.N.B., Kleinmann, S.G., and Ridgway, S.T. 1982, Astrophys. J., **253**, 136.

Shen, L.N., Doyle, T.J., and Graham, W.R.M. 1990, J. Chem. Phys., **93**, 1597.

Verhoeve, P., ter Meulen, J.J., Meerts, W.L., and Dymanus, A. 1986, Chem. Phys. Lett., **132**, 213.

Watson, D.M. 1984 in Galactic and Extra Galactic Infrared Spectroscopy, ed. Kessler, M.F., and Phillips, J.P., Reidel, pg. 195.

Werner, M. 1990, SIRTF Mirror; The Scientific Case for SIRTF, JPL 410-28-6.

Yamada, C., Hirota, E., Yamamoto, S., and Saito, S. 1988, J. Chem. Phys., **88**, 46.

Yamamoto, S. and Saito, S. 1991, Ap. J., **370**, L103.

# 7

# Atomic Data from the Opacity Project

C. Mendoza *

ABSTRACT. The "Opacity Project" is an international collaboration dedicated to the calculation of stellar envelope opacities and the accurate atomic data that are required. The volume and accuracy of the atomic database generated in the project are evaluated by making comparisons with the best available experimental and theoretical datasets. We also describe the main features of a database management system, referred to as TOPBASE, that has been developed for efficient use of the Opacity Project atomic data.

## 1. Introduction

The name "Opacity Project" (OP) refers to an international collaboration that was formed in 1984 to re-estimate stellar envelope opacities. The project has involved research groups from France, Germany, the United States, the United Kingdom and Venezuela. The approach adopted by the OP to compute opacities is based on a new formalism of the equation of state (see papers in the series "The equation of state for stellar envelopes" [1 - 4]) and on the mass production of accurate atomic data, as described in the series of papers "Atomic data for opacity calculations" (ADOC) [5 - 18]. A main challenge of the OP has therefore been to accomplish the systematic atomic data production by means of elaborate *ab initio* methods. This approach implies that on an individual scale the OP data accuracy would probably not surpass that of highly optimised and specific methods, but the atomic database that has been generated anticipates a statistical accuracy and completeness that have not been reached in the field before. General descriptions of the OP are given in [5, 19, 20, 21].

We have recently been involved in the process of validating the OP data, where detailed comparisons are being made with available experimental

---

* * IBM Venezuela Scientific Center, PO Box 64778, Caracas 1060A, Venezuela. Also, Visiting Fellow, IVIC, Venezuela.

and theoretical datasets in order to ensure data integrity. Furthermore, since the OP atomic data are likely to be used in other research fields beyond opacity calculations, we have been encouraged to develop a portable database management system (DBMS) to enhance their accessibility. A first prototype of this program, to be referred to as TOPBASE, has been implemented, and it is being used in the process of data checking. It is planned to distribute the complete package (OP database and its DBMS) in the near future.

The present review concentrates on a discussion of the atomic data that have emerged from this project, including an account of the comparative work that is in progress. We also describe the main features of the TOPBASE interface.

## 2. Numerical methods

Opacity calculations require oscillator strengths and photoionisation cross sections for transitions involving the ground states and excited states of astrophysically abundant ions. A systematic and complete treatment of these radiative processes is imperative due to the variety of thermodynamic scenarios (temperature, density and chemical mixtures) that are considered in stellar studies. The accuracy issue arises because the reliability of the standard opacity tables, computed with modest atomic physics approximations, has been criticised in the context of Cepheid models [22].

The numerical methods and approximations used to generate the OP atomic data are fully described in the ADOC series:

- I. General description [5]
- II. Computational methods [6]
- III. Oscillator strengths for C II [7]
- IV. Photoionisation cross sections for C II [8]
- V. Electron impact broadening of some C III lines [9]
- VI. Static dipole polarisabilities of the ground states of the helium sequence [10]
- VII. Energy levels, f-values and photoionisation cross sections for He-like ions [11]
- VIII. Line profile parameters for 42 transitions in Li-like and Be-like ions [12]
- IX. The lithium isoelectronic sequence [13]
- X. Oscillator strengths and photoionisation cross sections for O III [14]
- XI. The carbon isoelectronic sequence [15]
- XII. Line profile parameters for neutral atoms of He, N and O [16]
- XIII. Line profiles for transitions in hydrogenic ions [17]
- XIV. The beryllium sequence [18].

Further contributions to this series are in preparation. Papers in this series will hereafter be referred to as "ADOC-i", where "i" represents the ith member in roman numerals. In the following subsections we give a brief description of the key features of the numerical methods employed in the project.

## 2.1. The close-coupling approximation

Wavefunctions for ionic states, whether bound or free states, are calculated by numerical methods based on the *close-coupling* (CC) approximation of electron-ion collision theory [23]. In this approach the wavefunction for a system consisting of an electron in the field of an $N$-electron ion is expanded in terms of the ion eigenfunctions

$$\Psi^{SL\pi} = \mathcal{A}\sum_{i=1}^{I} \chi_i \frac{1}{r} F_i(r) + \sum_{j=1}^{J} c_j \Phi_j \ . \tag{1}$$

The ion is usually referred to as the *target* or *core*, and states of the total $(N+1)$-electron system are assigned the quantum numbers $SL\pi$ (total spin, orbital angular momentum and parity) if $LS$ coupling is assumed. $\mathcal{A}$ is the antisymmetrisation operator; $\chi_i$ is a vector coupled product of a target eigenfunction times functions for the spin and angle coordinates of the outer electron; $F_i(r)$ is the outer electron radial function and $\Phi_j$ is a bound-state type function for the entire system introduced to compensate for orthogonality conditions imposed on $F_i(r)$ and to improve short-range correlation. The electron radial function $F_i(r)$ and the coefficients $c_j$ are treated as variational parameters that lead to a set of integro-differential equations which are solved numerically. The finite number of terms $(i = 1, I)$ in the first expansion of equation (1) are referred to as *free channels*, while the correlation functions $(j = 1, J)$ in the second expansion are usually called *bound channels*.

The total energy of the system is

$$E = E_i + \varepsilon_i \ , \tag{2}$$

where $E_i$ and $\varepsilon_i$ are respectively the target and electron energies in the $i$th free channel. Channels with $\varepsilon_i \geq 0$ are referred to as *open channels* since $F_i(r)$ exhibits the oscillatory asymptotic behaviour characteristic of scattering states. For $\varepsilon_i < 0$, a *closed channel*, the electron radial function obeys the asymptotic boundary condition of a bound-state orbital

$$\lim_{r\to\infty} F_i(r) = 0 \qquad\qquad (\varepsilon_i < 0) \ . \tag{3}$$

A typical situation for a continuum state is to have open and closed channels, the interaction of which gives rise to the resonances observed in electron excitation and photoionisation cross sections of ionic systems. On the other hand, the case when all channels are closed takes the form of an eigenvalue problem for the $(N+1)$-electron system, where closed-channel coupling accounts for electron correlation effects (e.g., series perturbations) that are usually treated in structure calculations by the method of configuration interaction.

From the point of view of opacity calculations, the CC approach presents attractive advantages. Photoexcitation and photoionisation are treated in a consistent unified approach, whereby the same level of approximation is maintained for both bound and continuum states. This is particularly useful when complete spectral distributions of oscillator strength are required. The method can be made very efficient and can therefore be used to generate large amounts of radiative data. The accuracy of this approach, which is limited by the quality and number of target states included in the CC expansion, is comparable to that of the best methods that have been developed to study, usually independently, bound-bound and bound-free atomic transitions. Considerable effort is therefore devoted to obtaining accurate target representations using the configuration interaction codes SUPERSTRUCTURE [24] and CIV3 [25].

## 2.2. The R-matrix method

The numerical method used by the OP to solve the CC integro-differential equations implies dividing up the configuration space into an inner region ($r \leq a$, say), where all the detailed short-range electronic interactions are included, and an asymptotic region ($r \geq a$) where only long-range correlation is important. Solutions in the inner region are obtained by the *R-matrix* method [26], whereby a boundary condition is imposed at $r = a$ such that solutions, $\Psi = \psi_n$, only exist for a discrete set of energies $E = e_n$. This basis of "non-physical states", referred to as the *R-matrix states*, is then used to expand the wavefunction $\Psi_E$ (with radial part $F_{iE}(r)$) at any energy $E$

$$\Psi_E = \sum_n \psi_n A_{nE} \ . \tag{4}$$

The boundary condition is usually taken to be

$$f'_{in}(a) = 0 \ , \tag{5}$$

where $F_i(r) = f_{in}(r)$ is the radial part of $\psi_n$ and $y' \equiv dy/dr$ denotes the radial derivative. The expansion coefficients in equation (4) are obtained by diagonalising the Hamiltonian for the total system, and they can be expressed as

$$A_{nE} = (e_n - E)^{-1} \sum_i f_{in}(a) F'_{iE}(a) \ . \tag{6}$$

It may be shown that the radial function $F_{iE}(r)$ at the boundary radius is given by

$$F_{iE}(a) = \sum_{i'} R_{ii'}(E) F'_{i'E}(a) \ , \tag{7}$$

where

$$R_{ii'}(E) = \sum_{n} f_{in}(a)(e_n - E)^{-1} f_{i'n}(a) \tag{8}$$

is known as the *R-matrix*. One of the advantages of the R-matrix method is that the inner-region matrix diagonalisation is performed only once for each $SL\pi$ case.

"Physical states" are obtained by matching the solutions in the inner and outer regions at $r = a$ using expression (7), which in compact matrix notation can be written as

$$\mathbf{F} = \mathbf{RF}' \ . \tag{9}$$

In the outer region ($r \geq a$) bound channels can be neglected and the CC expansion takes the form

$$\Psi_{i'} = \mathcal{A} \sum_{i} \psi_i \frac{1}{r} P_{ii'}(r), \tag{10}$$

where $\Psi_{i'}$ denotes a set of linearly independent solutions. The **P** matrix is a solution of the ordinary differential equation

$$(h + \mathbf{v} - \varepsilon)\mathbf{P} = 0, \tag{11}$$

where $h$ is the diagonal matrix

$$h_{ii} = -\frac{d^2}{dr^2} + \frac{l_i(l_i+1)}{r^2} - \frac{2z}{r} \tag{12}$$

and **v** are long-range multipole potentials. It has been shown that, outside the R-matrix box, **v** can be handled by first-order perturbation theory within the Coulomb problem (ADOC-II).

In the case when all channels are closed, the conditions for matching at the boundary are

$$\mathbf{F} = \mathbf{PX}, \tag{13}$$

$$\text{and} \qquad \mathbf{F}' = \mathbf{P}'\mathbf{X}' \tag{14}$$

where **X** is a column vector. Substituting expressions (13) and (14) in equation (9) gives

$$(\mathbf{P} - \mathbf{RP}')\mathbf{X} = 0 \tag{15}$$

which takes the general form

$$\mathbf{B}(E)\mathbf{X} = 0 \ . \tag{16}$$

**Table 1.** Number of states and transitions for atoms considered in the OP.

| Z | States | Trans. |
|---|---|---|
| 1 | 55 | 330 |
| 2 | 108 | 645 |
| 3 | 134 | 798 |
| 4 | 232 | 1624 |
| 5 | 238 | 1744 |
| 6 | 815 | 9160 |
| 7 | 1083 | 13122 |
| 8 | 1391 | 18037 |
| 9 | 1716 | 18375 |
| 10 | 2009 | 32172 |
| 11 | 2141 | 37142 |
| 12 | 2459 | 46175 |
| 13 | 2774 | 54325 |
| 14 | 3251 | 65188 |
| 16 | 4109 | 88422 |
| 18 | 5289 | 126950 |
| 20 | 6450 | 182144 |
| 26 | 18632 | 911580 |
| | 52886 | 1607933 |

Equation (16) has solutions only at discrete energies that correspond to the eigenvalues of the $(N+1)$-electron system. A key feature of the OP approach is that systematic scans for eigenvalues have been implemented in the code by searching for zeros of the determinant of $\mathbf{B}(E)$; that is

$$\det[\mathbf{B}(E)] = 0 \ . \tag{17}$$

The resulting energies are then used as initial estimates to solve equation (16) by iterative methods. If $E_1$ is the energy of the lowest target state in the CC expansion, an *effective quantum number* $\nu$ can be defined by the relation

$$E = E_1 - z^2/\nu^2 \ . \tag{18}$$

The program makes a scan, at a predefined interval $\delta\nu$, for all bound states in the range

$$\nu_{min} < \nu < \nu_{max} , \tag{19}$$

where usually $\nu_{max} = 10$. Since this approach does not depend on initial experimental estimates, unmeasured states can be computed and therefore the issue of completeness is addressed.

## 3. OP atomic database

The OP has computed atomic data for all stages of ionisation of astrophysically abundant species: $Z = 1$ through $Z = 14$; $Z = 16$; $Z = 18$; $Z = 20$ and $Z = 26$. The data that will become available consist of:

(a) Identified term energies for states with $n \leq 10$, where $n$ is the active-electron principal quantum number, and active-electron orbital angular momentum $l \leq 3$ and in some cases $l \leq 4$. Term identifications are labeled with the notation $(Tnl; SL\pi)$ or, in the case of equivalent-electron states, with $(C; SL\pi)$, where $T$ refers to a target state and $C$ to an $(N+1)$-electron configuration.

(b) Oscillator strengths (f-values) for optically allowed transitions between all states considered in (a).

(c) Photoionisation cross sections of states in (a) that lie below the ionisation threshold. The cross sections have been computed with energy meshes fine enough to resolve the resonance structure.

The line broadening problem, essential in the calculation of Rosseland mean opacities, has been fully discussed in ADOC-V, ADOC-VIII, ADOC-XII and ADOC-XIII. Although electron collisional broadening can be satisfactorily treated within the CC approximation, empirical formulae are introduced due to the large number of transitions involved in the opacity calculations. Since line broadening data are therefore not contained in the OP database, which is the central topic of the present review, we will not discuss the problem further. In Table 1 the contents of the OP database are summarised; since the database is being periodically updated, the listed numbers are only representative. Seaton *et al* [21] give a comprehensive overview of the target approximations used for each isoelectronic sequence and the authors responsible for the calculations. Most ADOC papers provide detailed comparisons with previous work that highlight the main features of each isoelectronic sequence. Further comparisons are included in [27 - 37]. In the following sections we evaluate the OP term energies and configuration assignments, f-values, radiative lifetimes and photoionisation cross sections. The comparisons that are presented are mainly from the data checking process that is currently in progress.

## 4. Term energies

The OP has been involved in the production of non-relativistic term energies. Comparisons with other datasets will be carried out here in terms of *quantum defects*, $\mu(n)$, defined by the relation

$$E(Tnl) = E(T) - z^2/[n - \mu(n)]^2 , \qquad (20)$$

where $E(Tnl)$ is the term energy in Rydberg units, $E(T)$ is the energy of the corresponding series limit and $z$ is the effective charge of the ion. The

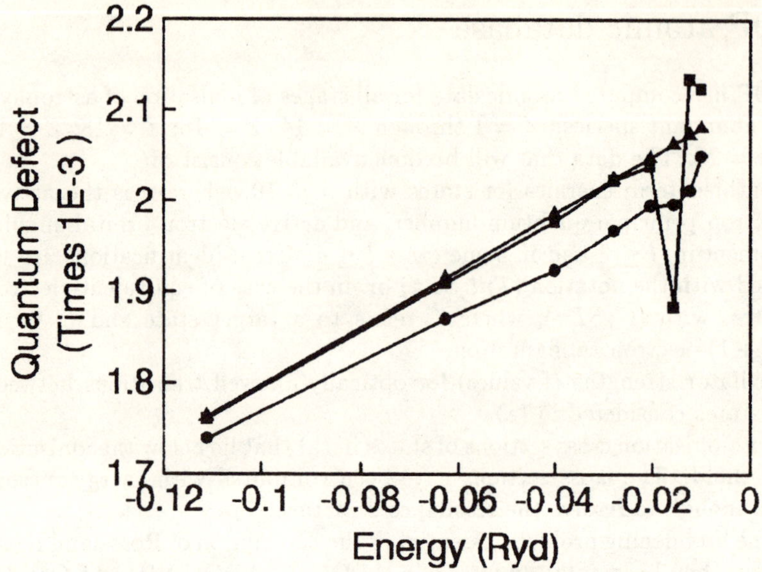


**Fig. 1.** Quantum defects for the $1snd\,^1D$ series of He I. Circles, OP results; squares, spectroscopic values [38], triangles, fitted experimental results [43 - 45]. Inaccurate spectroscopic values are observed for the highly excited states.

quantum defect gives a measure of the departure of a state from the hydrogenic energy position, and their comparisons and plots are particularly sensitive and bring out the finer features of spectroscopic series.

## 4.1. Comparison with experiment

We report here comparisons of OP quantum defects with spectroscopic measurements for ions of He, C, N, O and Si. The agreement with the spectroscopic tables of Martin [38] for He I is very good: for the 53 computed states the quantum defect r.m.s. deviation is 4%. This level of accuracy does not match, however, that obtained for the same ion by variational methods that include extensive Hylleraas-type basis functions [39 - 42]. In Fig. 1, OP and spectroscopic quantum defects for the $1snd\,^1D$ series of He I are plotted as a function of energy. Since the quantum defects for this system are expected to vary slowly with energy, the noticeable departures from such behaviour observed in the experimental values for the highly excited states are attributed to small errors in the measured term energies and/or series limit. The experimental error of the quantum defect,

$$\Delta\mu(n) = \frac{z}{2[E(T) - E(Tnl)]^{3/2}}[\Delta E(Tnl) + \Delta E(T)] , \qquad (21)$$

increases with $n$, and therefore becomes conspicuous for the highly excited states. A much higher level of accuracy can be obtained by introducing fitting techniques: the more consistent experimental results shown in Fig. 1 were obtained by fitting new measurements with the aid of reliable theoretical data [43 - 45].

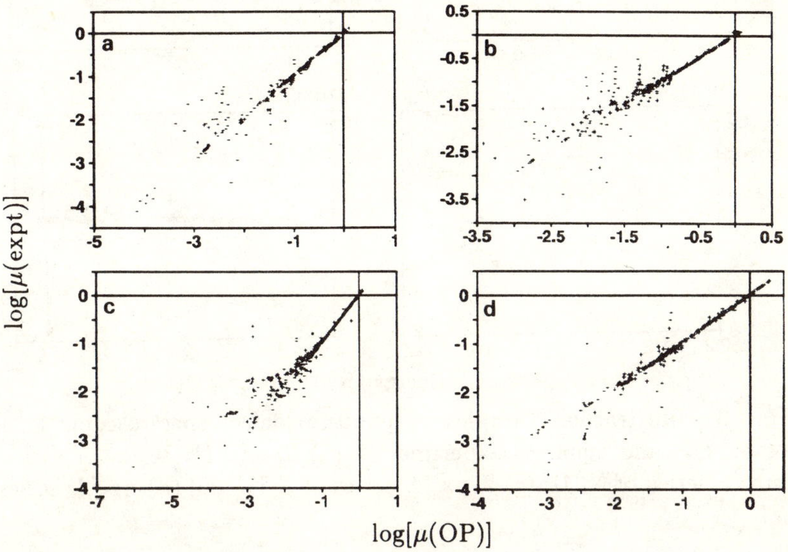


**Fig. 2.** Log plots of measured quantum defects against OP values for C, N, O and Si ions. (a) 317 states of C I - C V from [46]; (b) 467 states of N I - N VI [48 - 53]; (c) 482 states of O I - O VII [54 - 58]; (d) 491 states of Si I - Si XIII [59].

In Fig. 2 quantum defects for measured states of non-hydrogenic C, N, O and Si ions are compared with OP results. Several conclusions emerge from this comparison: (a) the ratio of the number of measured to computed levels decreases rapidly with $Z$, from $\sim$ 45% for C and N ions to 36% and 15% for O and Si ions respectively; (b) for distinctively non-hydrogenic bound states, i.e., for states with fairly large quantum defects, the agreement is very good and (c) differences for states with very small quantum defects can be large, but in these cases they are unimportant as such states can be treated as hydrogenic.

In Fig. 3 quantum-defect trends for representative spectroscopic series are depicted. The inaccuracies observed in the experimental excited states of O VI are similar to those discussed above in relation to He I. The peculiar behaviour of the measured quantum defects for the highly excited states of C I is due to breakdown of LS coupling [47] and, consequently, it is not reproduced by the OP results. The drastic effects caused by series

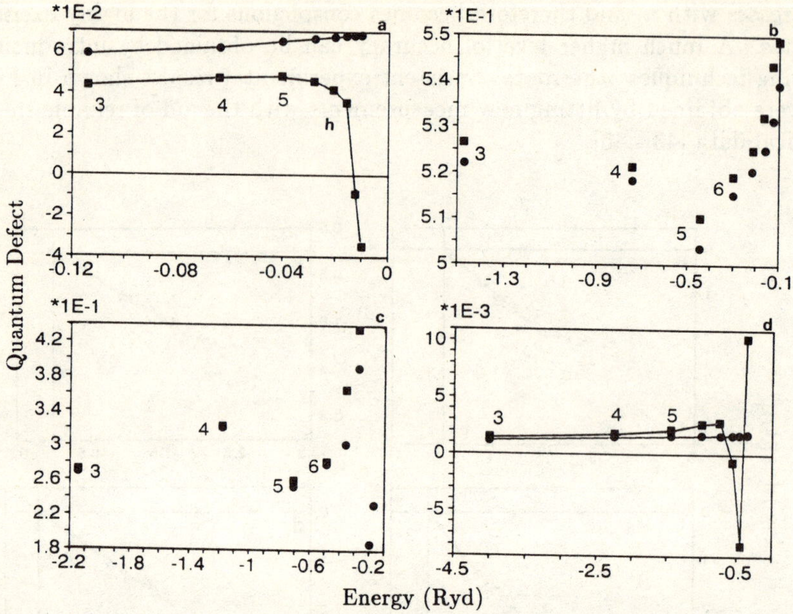


**Fig. 3.** Illustrative examples of quantum defect spectroscopic trends. Crosses, OP and squares, experiment. (a) $2pnd\ ^3D^0$ series of C I; (b) $2s^2ns\ ^2S$ series of N III; (c) $2sns\ ^1S$ series of N IV and (d) $nd\ ^2D$ series of O VI.

perturbers can be seen in the plots for the $2sns\ ^1S$ states of N IV and the $2s^2ns\ ^2S$ states of N III, where it may be appreciated how closely they are reproduced by the OP results. Term identifications in such situations can be a difficult task, particularly for strongly mixed states or when perturbers lie close to the series limit. For example, for strongly mixed states in N IV, such as the $2s8s\ ^1S$ and $2p4p\ ^1S$ or the $2s4f\ ^1F^0$ and $2p3d\ ^1F^0$ states, precise identifications may be of no practical importance; however, in the case of states belonging to the $^3S$ series, a comparison between the experimental and OP data exposes significant inconsistencies (see Table 2). Although the interchanged assignments for the $2s4s$ and $2p3p$ states are trivial, the spectroscopic term energy labeled $2s8s$ is questionable; furthermore, since a comparison with the theoretical energies suggests a reassignment of the experimental $2p4p$ term to $2s8s$, there does not seem to be a satisfactory measurement for the position of the $2p4p$ perturber.

## 4.2. Isoelectronic trends

A convenient way of analysing the isoelectronic trends of spectroscopic series is to plot the $z$ dependence of the effective quantum number (see, for instance, ADOC-XIV). The effective quantum number $\nu(T')$ for a state

**Table 2.** Comparison of experimental [48] and OP assignments for the $^3S$ series of N IV. It is shown that the experimental term energy labeled $2s8s$ is questionable and that the experimental $2p4p$ term should be reassigned to $2s8s$. Energies are given in Rydberg units relative to the ionisation threshold.

| State | $E_{op}$ | $\mu_{op}$ | $E_{expt}$ | $\mu_{expt}$ |
|---|---|---|---|---|
| 2s3s | −2.2536 | 0.3354 | −2.2562 | 0.3370 |
| 2p3p | −1.2455 | 0.1602 | −1.1557 | 0.0909 |
| 2s4s | −1.1518 | 0.2730 | −1.2508 | 0.4235 |
| 2s5s | −0.7288 | 0.3145 | −0.7295 | 0.3168 |
| 2s6s | −0.4961 | 0.3207 | −0.4965 | 0.3234 |
| 2s7s | −0.3623 | 0.3543 | −0.3635 | 0.3656 |
| 2p4p | −0.3251 | 0.1214 | −0.2695 | 0.0086 |
| 2s8s | −0.2691 | 0.2892 | −0.2496 | −.0071 |
| 2s9s | −0.2115 | 0.3021 | −0.2118 | 0.3090 |
| 2s10s | −0.1703 | 0.3061 | −0.1706 | 0.3164 |

assigned the $(Tnl; SL\pi)$ configuration is defined by the relation

$$E(Tnl; SL\pi) = E(T') - z^2/\nu^2(T'), \tag{22}$$

where $E(Tnl; SL\pi)$ is the term energy, $E(T')$ is the energy of the $T'$ core state and $z$ is the effective ionic charge. This definition implies that the effective quantum number depends on the reference core state $T'$. In Fig. 4, $\nu(2s)$ is plotted against $1/z$ for the $^3S$ states of the Be isoelectronic sequence. The intermixing of the $2sns$ and $2pnp$ series is clearly shown, particularly the perturbations caused by crossings that may be interpreted as "avoided crossings".

# 5. Oscillator strengths

In the next subsections, the agreement of OP f-values calculated in the length and velocity formulations is discussed, and we include some comments on the spectral distribution of oscillator strength. Comparisons with other datasets are also presented.

## 5.1. Comparison between length and velocity formulations

When elaborate methods are used to compute oscillator strengths, a comparison between results obtained in the length and velocity formulations is

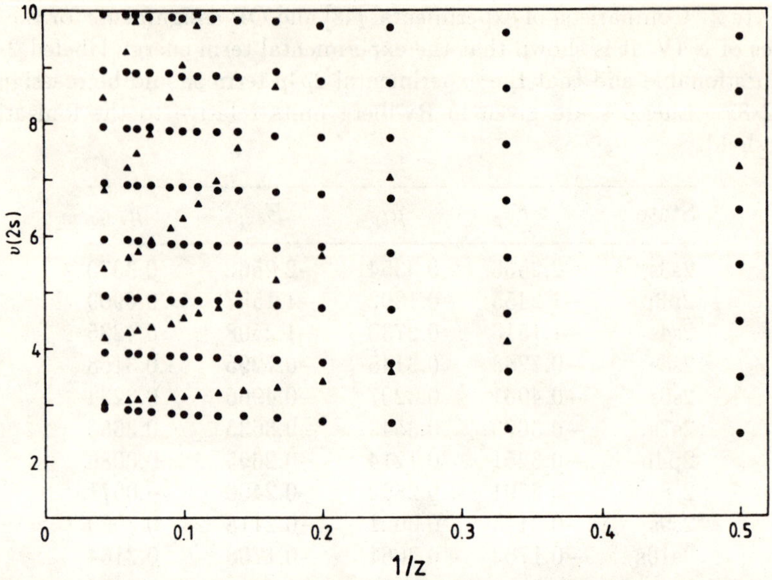


**Fig. 4.** Effective quantum number $\nu(2s)$ as a function of effective ionic charge $z$ for the $^3S$ series of the Be sequence. The intermixing of the $2sns$ (filled circles) and $2pnp$ (filled triangles) states is clearly shown.

usually given as a measure of the consistency and accuracy of the adopted approximations (for a discussion on the validity of the equivalence of the two forms, see Hibbert [60]). Since the OP f-value database typically contains a few thousand transitions for an ion, and taking into account that f-values may vary by orders of magnitude, an indicator of the level of correlation between the two formulations can be implemented in terms of an *average percentage difference* (ADOC-III). If two datasets are to be compared, $\{a\}$ and $\{b\}$ say, an average percentage difference is defined as

$$\Delta(a,b) = 100 \times \left[\sum_i (a_i - b_i)^2\right]^{1/2} \left[\sum_i a_i \times b_i\right]^{-1/2}. \tag{23}$$

For instance, it has been found that $\Delta(gf_l, gf_v) = 3.7\%$ for C II (ADOC-III), where $gf_l$ and $gf_v$ are the length and velocity weighted oscillator strengths respectively, and that $\Delta(gf_l, gf_v) \leq 3\%$ for ions in the Mg isoelectronic sequence [28]. We have extended this test to the complete database and find that $\Delta(gf_l, gf_v)$ is usually much less than 10% for most species, except for ions of Fe with electron numbers $N_e > 14$ where differences as large as 26% are found. This is due to relatively simple target representations and large numbers of poorly correlated $n = 3$ bound-channel states [29].

## 5.2. Spectral distribution of oscillator strength

A complete spectral distribution of oscillator strength is not only a requirement in opacity calculations, but it is useful when inspecting the radiative properties of atomic states. For instance, the spectral distribution of oscillator strength of a state $i$ obeys the sum rules

$$\sum_j f_{ij} + \int_0^\infty \frac{df}{d\epsilon} d\epsilon = N, \text{and} \tag{24}$$

$$\sum_j \frac{f_{ij}}{(E_j - E_i)^2} + \int_0^\infty \frac{1}{E^2} \frac{df}{d\epsilon} d\epsilon = \frac{\alpha}{4}, \tag{25}$$

where $N$ can be associated with the number of active electrons and $\alpha$ is the dipole polarisability of the state. (Here the polarisability is given in atomic units and the energies in Rydberg units.) These expressions allow consistency checks to be implemented on the radiative data and are indeed being used to examine the OP database. In Fig. 5, the oscillator strength distribution out of the $4s^2\ {}^1S$ ground state of Ca I is shown, where the anomalous behaviour in the discrete region is caused by a $3d4p\ {}^1P^0$ perturber. It can be seen that none of the calculations does well in this situation. The experimental distribution by Parkinson *et al* [63] has been extrapolated to the ionisation limit to obtain an estimate of 2.75 Mb for the the threshold photoionisation cross section. The agreement with theoretical threshold values in this situation is poor: 7.42 Mb (OP), 3.89 Mb [62], 3.5 Mb [65]. Experimental difficulties also contribute to discrepancies in measurements: 0.45 Mb [66], 0.90 Mb [67] and 1.25 Mb [68]. (Note: 1 Mb $\equiv$ $10^{-18}$cm$^2$.)

## 5.3. Comparison with theoretical datasets

Butler *et al* [28] have carried out comparisons of OP f-values for Mg-like ions with representative theoretical datasets. The results of these comparisons are shown in Fig. 6, where the number of transitions considered in each dataset and the corresponding average percentage differences are given. Since the OP database lists f-values for 17041 transitions in this sequence, the data volume increase over previous work is indeed significant. It can be seen that the accord with the MCHF and MS sets is particularly good.

We have also compared OP f-values with those listed for 157 optically allowed multiplet transitions in the recent compilation of atomic data for resonance absorption lines by Morton [79]. This comparison does not include all the transitions contained in this useful compilation since the latter also presents data for incomplete multiplet components and intercombination lines that were not considered in the OP. It is found that for about

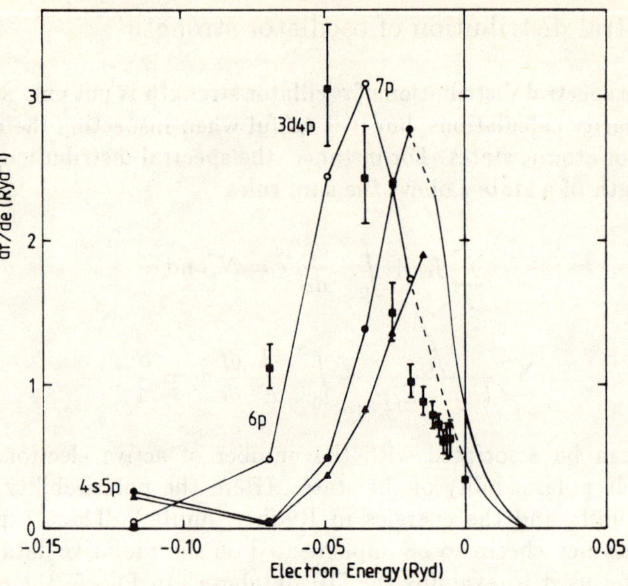


**Fig. 5.** Spectral distribution of oscillator strength (length formulation) of the $4s^2\ ^1S$ ground state of Ca I rendered by different methods. The strong transition to the $4s4p\ ^1P^0$ state is not shown. Filled circles, OP; filled triangles, multiconfiguration Hartree-Fock [61]; open circles, close-coupling with model potential [62]; filled squares, relative hook measurements normalised to an experimental lifetime [63]. The sensitive and anomalous distribution is caused by the $3d4p$ perturber that leads to poor agreement between theory and experiment.

half of the compared transitions the oscillator strengths agree to within 10%. Since most of the listed transitions belong to neutrals and lowly ionised species, where accurate calculations are the most difficult, this level of agreement can be considered satisfactory. However, there are transitions, most with small f-values, showing discrepancies larger than a factor of 2 that are briefly discussed below.

(a) Transitions $2p^2\ ^3P - 8s\ ^3P^0, -10s\ ^3P^0$ and $-7d\ ^3D^0$ in C I; $3p^2\ ^3P - nd\,^3D^0\ (n = 5, 7)$ and $-nd\,^3P^0\ (n = 5, 6)$ in Si I and $3p^4\,^3P{-}nd\,^3D^0\ (n = 5, 8)$ in S I. The values quoted by Morton were obtained from [80, 81], where no reliability is guaranteed for transitions in neutrals with small f-values.

(b) Transitions $4s^2\ ^1S - 4snp\ ^1P^0\ (n = 5, 7)$ and $-3d4p\ ^1P^0$ in Ca I. From the theoretical point of view, these are particularly difficult transitions, since the $4snp\ ^1P^0$ states are strongly perturbed by the $3d4p$ state. The listed values are taken from the compilation by Wiese *et al* [64] that are

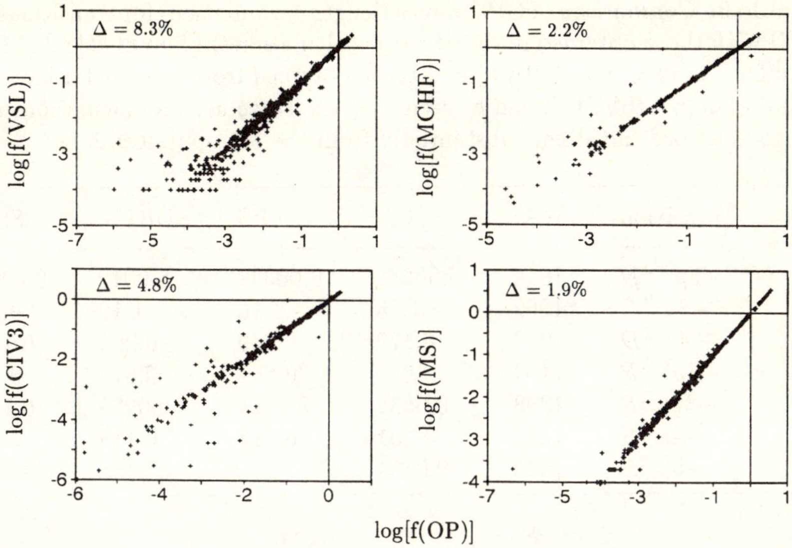


**Fig. 6.** Log plots of OP f-values (length formulation) for Mg-like ions versus those obtained by the best available theoretical methods. The average percentage difference ($\Delta$) for each comparison is quoted. VSL: 615 transitions from model potential calculations [69]. MCHF: Multiconfiguration Hartree-Fock results for 219 transitions [70 - 73]. CIV3: Configuration interaction calculations for 337 transitions [74 - 77]. MS: $L^2$ method (429 transitions) [78].

in good agreement with the most recent experiments and are probably more accurate than most theoretical results (see discussion in section 5.2).

(c) Transitions $2p^3\ ^4S^0 - 2p^4\ ^4P$ in N I and $3p^2\ ^3P - 4d\ ^3P^0$ and $-4d\ ^3D^0$ in Si I. The absolute values of the selected oscillator strengths for these transitions depend on estimates of radiative lifetimes. The accuracy of lifetime measurements is discussed in section 6 where we argue that in many cases it can be questionable.

(d) $2p^2\ ^3P - 7s\ ^3P^0$, $-9s\ ^3P^0$, $-5d\ ^3P^0$, $-6d\ ^3P^0$, $-9d\ ^3D^0$ and $-10d\ ^3D^0$ in C I and $3p\ ^2P^0 - 3p^2\ ^2S$, $-4s\ ^2S$, $-5s\ ^2S$, $-3p^2\ ^2D$ in Si II. The f-values included in the compilation for these transitions have been estimated from interstellar absorption lines. Such methods can give results that may substantially differ from those obtained by elaborate atomic calculations. For instance, in Table 3 we show that the OP f-values for transitions in Si II agree satisfactorily with results by other *ab initio* computations rather than with those obtained from astrophysical observations.

**Table 3.** Comparison of OP f-values (length formulation) for transitions in Si II with those obtained in the *ab initio* calculations of Luo *et al* [82] (LPS) and Dufton *et al* [83] (DHKT) and those obtained from interstellar lines by Shull *et al* [84] (SSY). It can be seen that the three atomic calculations are in good accord but differ substantially from the astrophysical data.

| Transition | $\lambda(\AA)$ | OP | LPS | DHKT | SSY |
|---|---|---|---|---|---|
| $3p\ ^2P^0 - 3p^2\ ^2D$ | 1814 | 0.00249 | 0.00347 | 0.0012 | 0.0055 |
| $-3d\ ^2D$ | 1263 | 1.18 | 1.16 | 1.12 | 0.96 |
| $-4d\ ^2D$ | 992 | 0.170 | 0.269 | 0.205 | 0.133 |
| $-3p^2\ ^2P$ | 1194 | 0.886 | 0.881 | 0.874 | 0.75 |
| $-3p^2\ ^2S$ | 1308 | 0.0885 | 0.0914 | 0.0929 | 0.147 |
| $-4s\ ^2S$ | 1531 | 0.131 | 0.118 | 0.119 | 0.23 |
| $-5s\ ^2S$ | 1023 | 0.0151 | | 0.0198 | 0.0283 |

## 6. Radiative lifetimes

The radiative lifetime of a state $j$ is defined as

$$\tau_j^{-1} = \sum_i A_{ji} \tag{26}$$

where $A_{ji}$ is the spontaneous decay transition probability. The latter can be expressed in terms of the absorption oscillator strength by

$$A_{ji} = \frac{\alpha^3 E_{ij}^2}{2\tau_0} \frac{g_i}{g_j} f_{ij} \tag{27}$$

where $\alpha$ is the fine structure constant, $\tau_0$ is the atomic time unit, $E_{ij}$ is the photon energy in Rydberg units and $g_i$ and $g_j$ are statistical weights.

The comparison between calculated and experimental radiative lifetimes has revealed difficulties on both sides. In some experimental methods, e.g., beam-foil spectroscopy, the excitation of the $j$th state is carried out by non-selective methods that can lead to cascade effects. The accurate determination of cascade contributions in such experiments has only been accomplished fairly recently (see the review by Martinson [85]). However, interesting new approaches have been proposed to attain selective excitation using a combination of the beam-foil technique and lasers that ensure cascade-free measurements (see, for instance, [86, 87]). Developments in instrumentation have also led to improvements when dealing with line blends and ion velocity measurements, consequently reducing experimental uncertainties. Thus, although lifetime measurements have suffered from systematic errors in the past and comparisons with theoretical results have

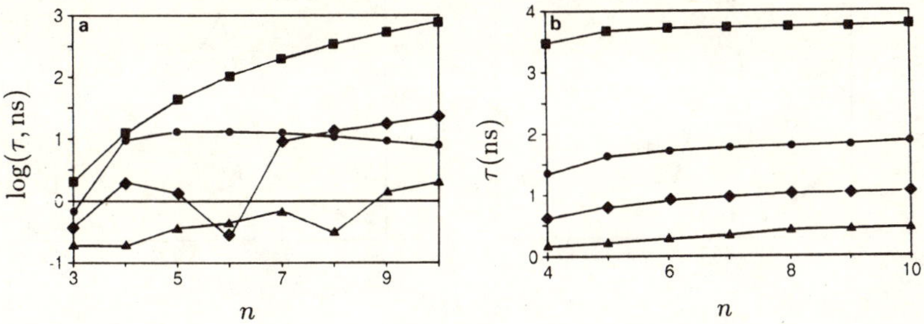


**Fig. 7.** Radiative lifetimes (ns) from [33] as a function of $n$ for (a) the $3snp\ ^1P^0$ and (b) $3pnp\ ^1P$ states of Mg I (squares), Al II (crosses), Si III (diamonds) and S V (triangles). A marked increase in the lifetimes of the $3snp\ ^1P^0$ states with $n$ is seen in spite of the effects of series perturbers. In contrast, the lifetimes of the $3pnp\ ^1P$ doubly excited states are found to be practically independent of $n$. The latter effect is caused by the dominance of the channel that involves the photo-decay of the core.

not always been encouraging, the more recent work should produce accurate data. This point is illustrated in a recent compilation of f-values and lifetimes for C, N and O ions by Allard *et al* [32].

Theoretical work, on the other hand, is concerned with the production of f-values (transition probabilities) that must be summed up over all possible decay channels in order to obtain lifetimes (see equation 26). Important correlation effects can lead to large involved computations, particularly when treating highly excited states by the method of configuration interaction. Since the number of decay channels increases dramatically for the excited states, in most cases there are not sufficient spectroscopic data to be able to determine the complete decay manifold, and therefore methods that depend on initial experimental energy estimates can only offer lifetime upper bounds.

The completeness of the OP f-value database allows the comparison of fully converged theoretical lifetimes with experiment for most ionic states, and it provides an incentive for the study of isoelectronic and spectroscopic trends. In the following subsections these capabilities are illustrated with a few interesting examples.

## 6.1. Spectroscopic and isoelectronic trends

We briefly describe here the work carried out in [33] on radiative lifetimes for states in Mg-like ions. In Fig. 7 we plot the lifetimes of the $3snp\ ^1P^0$ and $3pnp\ ^1P$ states of Mg I, Al II, Si III and S V as a function of $n$. The

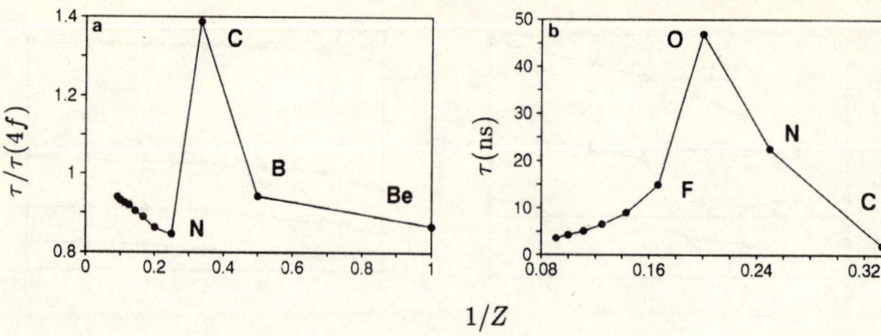


**Fig. 8.** Lifetime isoelectronic trends from the OP for (a) the $2s4f\ ^3F^0$ state and (b) the $2p3d\ ^3F^0$ perturber in the Be sequence. The lifetime of the $2s4f$ state is plotted relative to the hydrogenic $4f$ in order to show the enhancement produced before the crossing, which occurs between C III and N IV. The lifetime of the $2p3d$ state is also greatly increased, but the peak is observed after the crossing. Lifetimes are given in ns.

lifetimes of the singly excited $3snp\ ^1P^0$ states increase with $n$, but the effects caused by "plunging" perturbers are clearly seen. In the case of Al II, the effect is not localised since there has not been a crossing, while the lifetime of the $3s6p$ state of Si III shows a large decrease due to a strong admixture with the $3p3d$ perturber. In contrast, the lifetimes of the doubly excited $3pnp\ ^1P$ states (these states lie above the ionisation threshold but do not autoionise in LS coupling) are almost independent of $n$ and the effect of perturbers is not noticed. This behaviour is due to the fact that the radiative decay process is dominated by the channel

$$3pnp\ ^1P \rightarrow 3snp\ ^1P^0 + \gamma \tag{28}$$

that corresponds to a photo-decay of the core, where the active electron behaves as a spectator. This process may be regarded as the reverse of the photoexcitation of core states (PEC) process (see subsection 7.1).

The effects on lifetimes caused by the crossing of a perturber can also be characterised by studying their $z$ dependence. In Fig. 8 we plot the OP lifetimes of the $2s4f\ ^3F^0$ and the $2p3d\ ^3F^0$ states as a function of the effective ionic charge $z$. The lifetime of the $2s4f$ states is shown to be almost hydrogenic, except when the mixing with the $2p3d$ leads to enhancements as large as 50%. The effect of the crossing, which takes place between C III and N IV, also causes a large increase in the lifetime of the perturber.

## 6.2. Comparison between theory and experiment

In Fig. 9 we compare OP lifetimes for He I and Mg-like ions with experiment and other theoretical results. It can be seen that the correlation of the

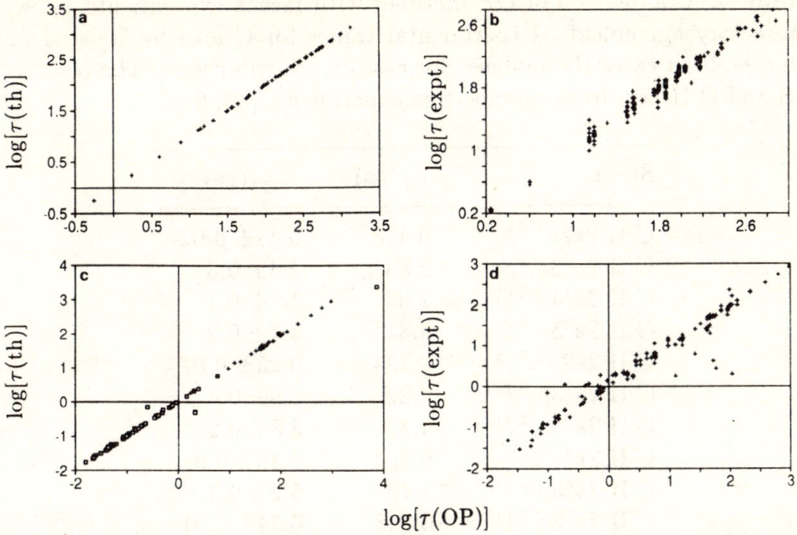


**Fig. 9.** Comparison of OP lifetimes for He I and Mg-like ions with experiment and theory. (a) Calculated values for He I by Theodosiou [88]. (b) Experimental lifetimes for He I compiled in [88]. (c) Theoretical results for ions in the Mg sequence from [73, 78]. (d) Measurements for Mg-like ions compiled in [33]. Lifetimes are given in ns.

OP data with other theoretical results is at much higher level than with experiment. The r.m.s. deviations with the theoretical sets are 2% for He I and 8% for Mg-like ions. In the comparison with Mg-like ions we have excluded the very long-lived $3p^2\ ^1D$ state of Al II and the strongly mixed $3p3d\ ^1P^0$ and $3s6p\ ^1P^0$ states of Si III where expected large differences are encountered. On the other hand, the r.m.s. deviations with experiment are 19% and 22% for He I and Mg-like ions, respectively. (In the latter set, we have excluded a few states with discrepancies, which are most likely due to experimental problems, larger than a factor of 2.)

The agreement of OP lifetimes with recent experiments where cascade contributions are extensively analysed or eliminated (see Table 4) is found to be more encouraging (r.m.s.=10%). This comparison provides evidence in support of our earlier comments on the reliability of the older measured lifetimes.

# 7. Photoionisation cross sections

Total photoionisation cross sections have been calculated in the OP for all bound states lying below the ionisation threshold. Cross sections for LS

**Table 4.** Comparison of OP lifetimes with recent experiments, showing satisfactory agreement. Experimental values for C ions by Reistad *et al* [89] have been carefully analysed for cascade contributions. The results for N II and O III are from cascade-free experiments [86, 87].

| State | $\tau_{op}$(ns) | $\tau_{expt}$(ns) |
|---|---|---|
| C II $2s2p^2\ ^2S$ | 0.418 | 0.44± 0.02 |
| C II $2s^23s\ ^2S$ | 2.83 | 2.4± 0.3 |
| C II $2s^24s\ ^2S$ | 2.04 | 1.9± 0.1 |
| C II $2s^25s\ ^2S$ | 3.81 | 3.7± 0.2 |
| C II $2s2p^2\ ^2P$ | 0.234 | 0.25± 0.01 |
| C II $2s^23p\ ^2P^0$ | 9.22 | 8.9± 0.4 |
| C II $2s^24p\ ^2P^0$ | 4.36 | 3.8± 0.2 |
| C II $2p^3\ ^2P^0$ | 0.504 | 0.48± 0.02 |
| C II $2s^25p\ ^2P^0$ | 4.79 | 5.2± 0.3 |
| C II $2s^23d\ ^2D$ | 0.348 | 0.34± 0.01 |
| C II $2s^24d\ ^2D$ | 0.723 | 0.75± 0.03 |
| C III $2s2p\ ^1P^0$ | 0.532 | 0.57± 0.02 |
| C III $2p^2\ ^1S$ | 0.417 | 0.51± 0.03 |
| C III $2p^2\ ^1D$ | 7.45 | 7.2± 0.2 |
| C III $2s3s\ ^1S$ | 1.35 | 1.17± 0.05 |
| N II $2p3s\ ^1P^0$ | 0.253 | 0.267± 0.010 |
| N II $2p3d\ ^1F^0$ | 0.266 | 0.249± 0.004 |
| O III $2p3s\ ^1P^0$ | 0.215 | 0.17± 0.01 |

bound states with term energies above threshold are discussed in ADOC-IV.

Cross sections are computed with subtle energy meshes chosen to resolve the resonance structure. If a $z^2$-scaled energy, $\mathcal{E}$, is defined with respect to the ionisation threshold, the mesh is defined by the parameters

$$\mathcal{E}_{min}, \mathcal{E}_{max}, \nu_{max}, \delta\nu \quad \text{and} \quad \delta\mathcal{E}_{open} \tag{29}$$

with points being obtained in the range

$$\mathcal{E}_{min} \leq \mathcal{E} \leq \mathcal{E}_{max} \ . \tag{30}$$

In the resonance region, i.e., where some channels are closed, an effective quantum number $\nu_i$ can be defined by

$$\mathcal{E} = E_i/z^2 - 1/\nu_i^2 \ . \tag{31}$$

If $i$ is the lowest target state giving $\nu_i < \nu_{max}$, the interval in $\mathcal{E}$ is taken to be such that $\delta\nu_i = \delta\nu$. This furnishes equal resolution to all resonances converging to $E_i$ up to $\nu_i = \nu_{max}$. In the region just below a threshold

$$E_i/z^2 - 1/\nu_{max}^2 < \mathcal{E} < E_i/z^2 \ , \tag{32}$$

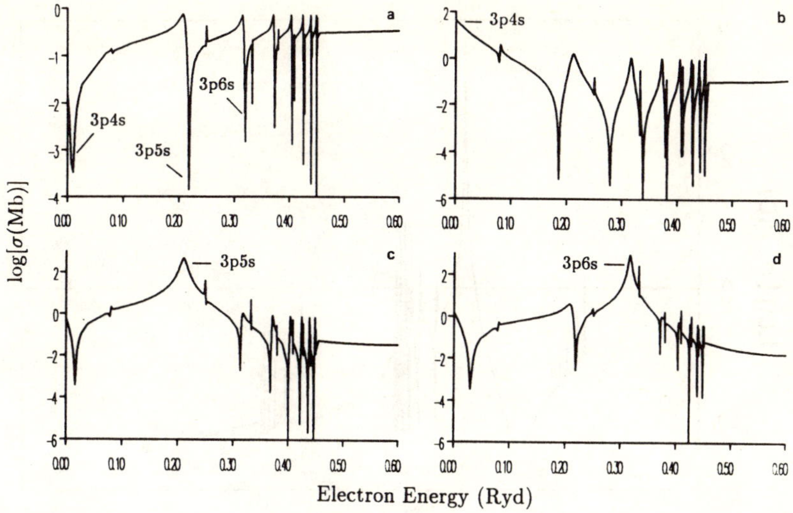


**Fig. 10.** OP photoionisation cross sections of the $3sns\ ^1S$ series of Al II. a) $3s^2$, b) 3s4s, c) 3s5s and d) 3s6s. Broad PEC resonances are seen in the cross sections of the excited states.

with an infinite number of resonances, the cross sections are given averaged over resonances (ADOC-IV). Finally, in the region where all channels are open, a fixed interval, $\delta\mathcal{E} = \delta\mathcal{E}_{open}$, is chosen. Cross sections calculated with these energy meshes ensure complete spectral distributions of oscillator strength in the continuum to complement the discrete spectrum.

The OP approach allows, for the first time, the systematic study of the photoionisation of excited states, thus exposing new effects such as PEC resonances that are important in opacity calculations. In the next subsections we discuss these effects and provide comparisons with experiment and theory.

## 7.1. PEC resonances

PEC resonances are broad features found in the photoionisation cross sections of specific excited states, where they stand out from the rest of the resonance structure by as much as orders of magnitude. They are produced, and thus their name, by the photoexcitation of a core state where the active electron takes the role of a spectator.

To illustrate this process we consider the photoionisation cross sections of the $3sns\ ^1S$ states of Al II with $n = 3to6$ (see Fig. 10). The cross sections are dominated below the first excitation threshold ($3p\ ^2P^0$) by a

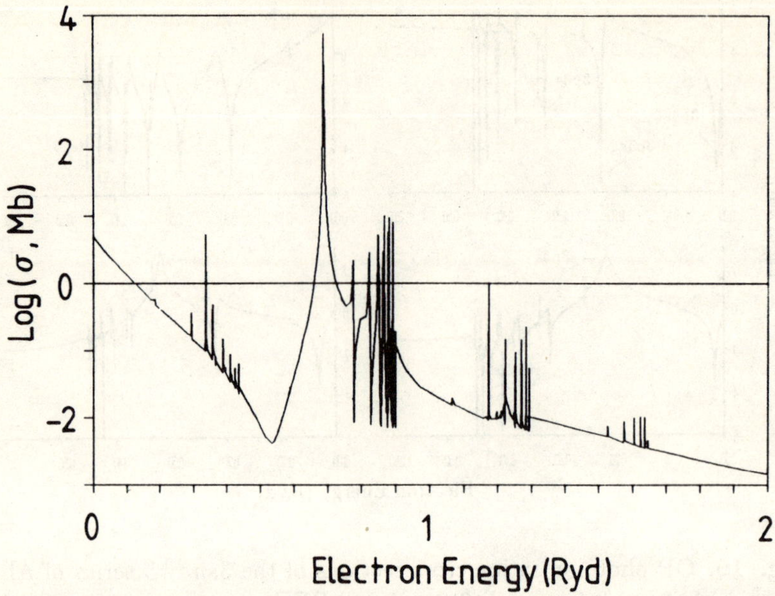


**Fig. 11.** Photoionisation cross section of the $2s^24f\ ^2F^0$ state of C II. This cross section, expected to be close to a hydrogenic cross section (He II 4f), is dominated at low energies by the $2s2p4f$ PEC resonance.

series of window resonances due to transitions of the type

$$3sns\ ^1S + \gamma \rightarrow 3pn's\ ^1P^0 \rightarrow 3s\ ^2S + e\ . \quad (33)$$

For example, the $3p4s\ ^1P^0$ resonance causes the cross section of the $3s^2\ ^1S$ ground state to be very small at threshold (see Fig. 10). PEC resonances arise for the cases when $n = n'$ in equation (33), and they can be seen in the cross sections of the excited states: the resonances for which this condition is satisfied turn from window resonances to peaks several orders of magnitude larger.

Another interesting case is that shown in ADOC-IV which relates to the noticeable effects caused by a PEC resonance on the cross section of a nearly hydrogenic state, namely the $2s^24f\ ^2F^0$ state of C II (see Fig. 11). The cross section at low energies can be seen to be dominated by the broad $2s2p4f$ PEC resonance.

## 7.2. Comparison with experiment

Absolute measurements of photoionisation cross sections are mainly performed on the ground state of neutrals, although there are some experiments on excited states based on techniques such as two-step laser excitation [90]. In Fig. 12 we compare OP cross sections with some experimental results. The agreement is found to be very good except in the high-energy region of the ground-state cross sections of Na I and Ar I. The discrepancy in the cross section of Na I is believed to be due, after much theoretical attention, to experimental error (see, for instance, Dasgupta and Bhatia [91]). An interesting case is the cross section of the $3s3p\ ^1P^0$ excited state of Mg I (see Fig. 12). As shown, the cross section in the near-threshold region is dominated by a broad resonance identified with the $3p^2\ ^1S$ state. The OP position and width for this resonance are in poor agreement with experiment, and previous CC calculations have not been able to match the experimental parameters (see discussions in [92, 93]). The cause of this discrepancy is not altogether clear.

### 7.3. Comparison with theory

In spite of many previous theoretical results on photoionisation, most approaches have been piecemeal. For this reason we only compare here with the work by Reilman and Manson [101], since they present a systematic treatment of photoionisation of the ground states of positive ions and their tables are frequently used by the astronomy community. In Fig. 13 we compare selected OP cross sections with their data. It can be seen that they do not obtain the resonance structures due to the neglect of correlation effects, but their results agree fairly well with the OP background cross sections. The agreement begins to deteriorate for low stages of ionisation and neutrals; for the case of O I (see Fig. 13), the differences in the near-threshold region are significant.

## 8. Conclusions

Before proceeding with the description of the TOPBASE DBMS, it is useful to draw some conclusions from the above discussion of the OP atomic database. We have attempted to show that the volume of LS data computed in the project surpasses those of spectroscopic tables, theoretical datasets or atomic data compilations by remarkable amounts. The OP data accuracy is in general comparable with the state of the art in atomic calculations, and the comparison with measurements is in most situations very encouraging. The examples that were reviewed in this paper were taken from a detailed data validation effort that is currently in progress to ensure data integrity, locate poor data and numerical difficulties and, when appropriate, to suggest improvements and recalculations. Finally, the completeness and extension of the OP atomic database are allowing the study of spectroscopic and isoelectronic trends of atomic properties in ways that were not possible before. The new effects that were discussed here became apparent during the actual computation of the data; perhaps many more will be exposed with further data processing.

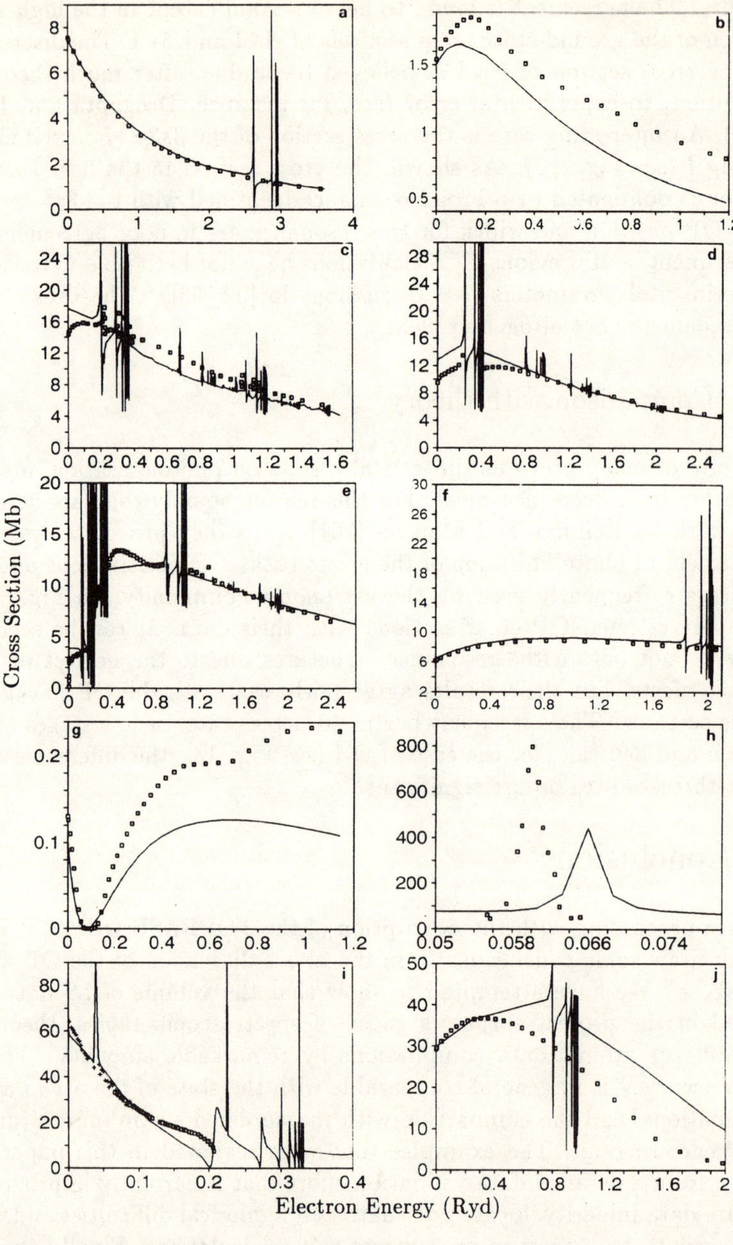


**Fig. 12.** Comparison of OP photoionisation cross sections (solid curve) with experiment (squares and crosses). a) He I ground state [94]; b) Ne I ground state [94]; c) Ar I ground state [94]; (d) C I ground state [95]; (e) N I ground state [96]; (f) O I ground state [97]; (g) Na I ground state [98]; (h) $3s3p\ ^1P^0$ excited state of Mg I [90]; (i) Al I ground state [99, 100].

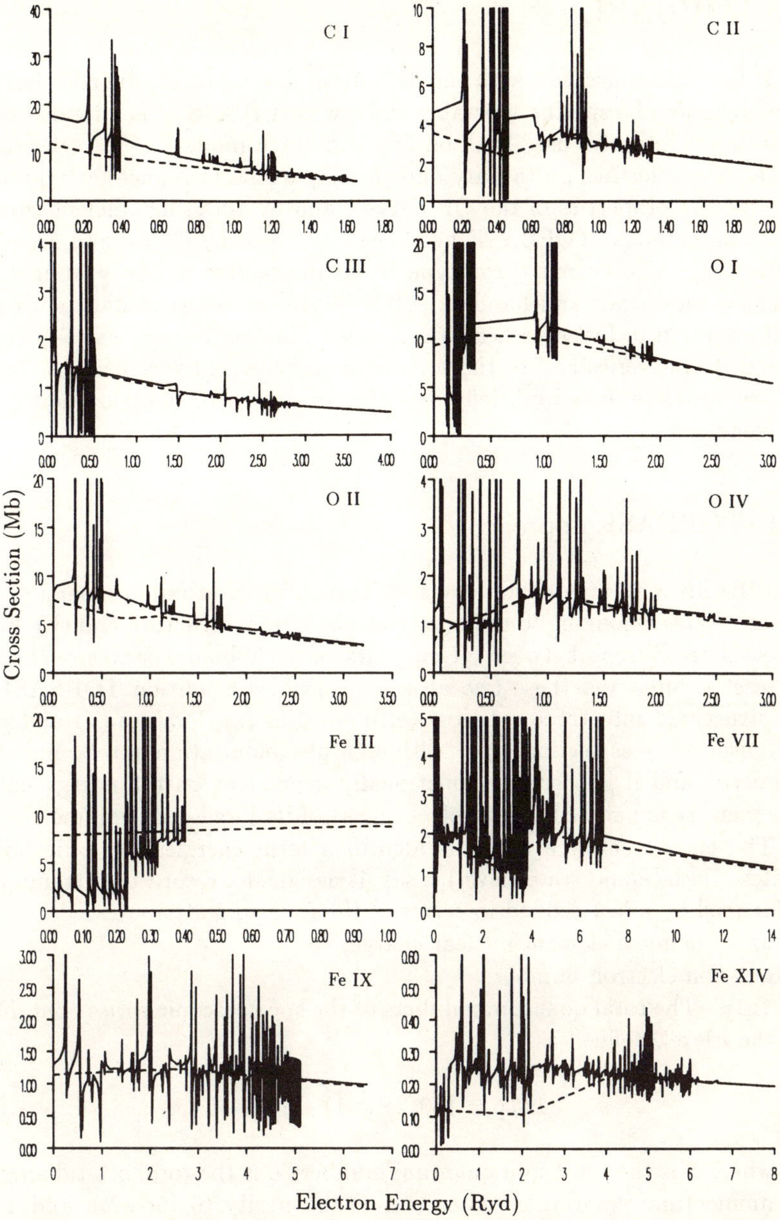


**Fig. 13.** Comparison of OP photoionisation cross sections for selected ground states (solid curve) with the theoretical data of Reilman and Manson [101] (broken lines). It can be seen that the latter work does not obtain the resonance structure due the neglect of correlation effects, but the background cross sections are in good agreement except perhaps for neutrals and lowly ionised species.

# 9. TOPBASE

TOPBASE is concerned with the definition, development, distribution and maintenance of a specific, portable and low-cost DBMS; it is intended to facilitate the intensive use –both on-line and batch modes– of the OP atomic data. It emphasises portability and time/space performance by exploiting the specific properties of the OP dataset and by adopting efficient computational methods. TOPBASE has been developed by Cunto and Mendoza [102, 103]. The current prototype is for interactive use only, it includes graphic commands and handles $\sim 0.5$ Gbytes of compact data with special attention to fast searches along isoelectronic and isonuclear sequences, spectroscopic series and to the sorting of energies and wavelengths. In the following subsections we briefly describe its approach, structure and query language.

## 9.1. TOPBASE approach

TOPBASE manipulates data related to two main atomic properties: (a) bound states belonging to the non-relativistic term structure and (b) dipole allowed transitions between given pairs of such bound states. The OP dataset volume and these two atomic properties condition TOPBASE to be structured into three entities: term energies ($e$), f-values ($f$) and photoionisation cross sections ($p$). Although photoionisation is a bound-state property, and it is therefore functionally dependent on the energy entity, the $p$ entity is handled separately because of its large data volume.

The energy entity $e$ contains identified term energies for ionic bound states. Each bound state (level), $i$ say, is assigned a record that is uniquely addressed by a key defined in terms of the following attributes:

- **nz** - Chemical element nuclear charge;
- **ne** - Ion electron number;
- **islp** - The total quantum numbers of the spectroscopic series containing the $i$ level, defined as

$$\texttt{islp} = 100(2S+1) + 10L + P, \tag{34}$$

  where $S$ is the total spin quantum number, $L$ is the total orbital angular momentum quantum number and $P$ the parity (0 for even and 1 for odd);
- **ilv** - The $i$ level energy position within its spectroscopic series.

Indexing in the $e$ entity is organised such that a hierarchy of access is enforced. This means that a single element with nuclear charge **nz** gives rise to `nz` ions with `ne=1,nz` electrons; each ion has several spectroscopic series defined by their total quantum numbers `islp`; and each series contains several levels labeled by `ilv` in ascending energy order.

The *f* entity contains weighted oscillator strengths (gf-values) and wavelengths for allowed transitions between pairs of bound states, $i \rightarrow j$ say, contained in the *e* entity. Each transition is uniquely addressed by a key defined by the attributes:

- `nz` - Chemical element nuclear charge
- `ne` - Ion electron number
- `islp` - The total quantum numbers of the spectroscopic series containing the *i* level
- `jslp` - The total quantum numbers of the spectroscopic series containing the *j* level
- `ilv` - The *i*-level energy position within its spectroscopic series
- `jlv` - The *j*-level energy position within its spectroscopic series.

Indexing in the *f* entity imposes a similar access hierarchy to that described for the *e* entity.

The photoionisation entity *p* may be regarded as an addendum of the *e* entity. Both entities share the same accessing scheme, that is, each state key identifies a unique photoionisation cross section. Since the cross section is a complicated function of photoelectron energy, the number of points varies among states, and therefore the *p* entity is voluminous and embodies $\sim 90\%$ of the OP dataset. Thus, for both convenience and performance, the *e* and *p* entities are kept physically separated although they are logically closely related.

The design framework takes into consideration the types of queries most likely to be performed by users. Let us consider, for example, queries addressing the *e* entity (the three entities are accessed separately). If each key attribute is assigned a value, or a range of values specified in terms of its lower and upper bounds, the query is referred to as *isonuclear* if the nuclear charge is kept constant at `nz=nz1` and the ion electron number is given a range `ne=ne1,ne2`. Similarly, if the number of ion electrons is assigned a fixed value `ne=ne1` and the nuclear charge is given a range `nz=nz1,nz2`, the query is said to be of the *isoelectronic* type. If `nz=nz1`, `ne=ne1` and `islp=islp1,islp2`, the query addresses multiple data belonging to a single ion, and it may be called an *ionic* query. In the case of a query with `nz=nz1`, `ne=ne1`, `islp=islp1` and `ilv=ilv1,ilv2`, data within a single spectroscopic series are accessed, i.e., a *spectroscopic* query. Finally, the user may be interested in states within an energy range `e=e1,e2`; this query is then referred to as an *energy* query. When searches are performed in the *f* entity, the query types are basically the same as those mentioned, except that the energy query becomes a *wavelength* query. Queries may indeed get more complicated, but the majority can be reduced to combinations of the above types.

Since data access is on a read-only basis, specific compact and efficient data structures have been devised to implement indices and tables. Access-time efficiency is ensured through a physical design tailored to economise in the number of data transfers to main memory from the bulk of the database that resides in secondary storage. Data loading from secondary storage is really the main bottle-neck in system performance due to the volume of data that may be involved. This problem becomes eminent in smaller platforms, e.g., stand-alone workstations with large memories and fast CPUs but relatively poor disk access times. For these reasons, the TOPBASE data manipulation scheme has been implemented at two levels: (i) searches in and expensive block data retrieval from secondary storage and (ii) cheap and versatile functions (sorting, column/row selections and exclusions, and graphic displays) performed iteratively on the data in main storage in order to satisfy the user's ultimate requirements.

## 9.2. General structure and data manipulation scheme

The data are organised in three entities, the $e$, $f$ and $p$ entities, and a set of indices all resident in secondary storage. When TOPBASE is invoked, the indices are loaded into main memory such that the query processor is ready to use them for searching the required data in the bulk of the database. Indices have been carefully structured so as to: (a) reduce a single search to one disk access when performed in the $e$ and $f$ entities and to two disk accesses when it is carried out in the $p$ entity; (b) optimise multiple searches assuming the query types described in section 9.1 and (c) display tables of contents of the database.

Data compactness and fast access are two main features of TOPBASE. In order to meet such requirements, physical media such as main and secondary storage are managed jointly with the logical handling of the data. Two data structures in main storage have been implemented for this purpose: the *view* and the *table*. A search in the database is performed according to user selected criteria that generate a subset of highly cohesive data sharing a common meaning. This subset is loaded into special buffers located in main memory to allow further manipulation. The data structure implemented in main memory to store the loaded subset is referred to as the *view*. Each view has an associated *descriptor* that registers the selection criteria and view bounds. Note that the basic idea of the view is to perform only one expensive search operation in secondary storage, and from that point on to facilitate inexpensive related queries on the highly cohesive data subset in main storage. Binary images of views can be archived in and restored from secondary storage. The software package offers display facilities on different output devices, namely the monitor, printer and disk files.

Logical reorganisations of the data stored in a view are possible through the concept of *tables*. A table is a vector array that enables or disables data items within the view, according to selection criteria, inclusion/exclusion facilities and sorting. Tables can be output on several devices (monitor, printer, disk files), and graphic displays of table columns and photoionisation cross sections can also be rendered.

## 9.3. Query language

Since TOPBASE is destined to be used on different platforms and from a large variety of terminals, its on-line user interface has been implemented as a command interpreter that recognises a simple although powerful query language.

A command consists of a `<command_name>`, usually followed by one or more `<arguments>` and, in most cases, by an `<option>` list. The command format takes the form

`<command>` ⟶ `<command_name>` `<arguments>` `<options>` .

The `<command_name>` is an alphanumeric symbol based on a verb that describes the function the user wants the DBMS to perform. The present version offers five types of commands:

- *View* commands: perform functions on the view data structure
- *Table* commands: perform functions on the table data structure
- *Graphic* commands: display graphic output of tables and photoionisation cross sections
- *Index* commands: provide tables of contents of the database
- *Constant* commands: list values of frequently used atomic constants.

The `<arguments>` are positional operands that specify the information on which the system operates when it performs a command. They can be an entity ($e$, $f$ or $p$), a pair of row limits, or a list of entity attributes and entity functions. The `<options>` are sequences of attribute items used to specify the selection criteria required in an operation, the execution control of the command or for input/output routing. Each attribute item is specified by an attribute identifier and an attribute range. The latter may be given a single value or a range of values; when an attribute is omitted, or its range is left undefined, the whole attribute range is assumed. In the present version of TOPBASE there are four types of option lists.

- `<descriptor_options>`: they specify the atomic data subset required when a view is created and take the form of a list of key attribute ranges. The view descriptor is updated with the option ranges at every view creation. This option list is preceded by a left parenthesis, "(".
- `<selector_options>`: they select the view subset when tables are created. They share exactly the same format as `<descriptor_options>` but do not lead to updates of the view descriptor.

- `<graphic_options>`: they provide specifications required in graphic commands, e.g., number of plots/screen, scale ranges and types, headers, etc. This option list is also preceded by a left parenthesis, "(".
- `<input/output_options>`: they determine the I/O device required by a command. The input and output options are preceded by the "<" and ">" indicators respectively.

Detailed descriptions of the TOPBASE commands are contained in its user's manual [102].

## 9.4. Query example

A frequent task in atomic physics is to plot quantum defects for states within a spectroscopic series when comparing with experiment and to estimate numerical accuracy and the effect of series perturbers. In order to illustrate the TOPBASE features, we have chosen for this example the $np\ ^2\mathrm{P}^0$ series of the astrophysically important C II ion. The first operation is to access the energy entity *e* and to load into memory (i.e., create a view) the $^2\mathrm{P}^0$ states of C II . This is accomplished by issuing the command

```
cv e (nz=6 ne=5 islp=211
```

Note that the `islp=211` is determined by the relation

$$\mathtt{islp} = 100(2S+1) + 10L + P\ , \tag{35}$$

where $S = 1/2$ is the total spin quantum number, $L = 1$ is the total orbital angular momentum quantum number and $P = 1$ denotes odd parity. As a result, 11 states are loaded from disk into the view buffer. Their attributes may be displayed on the screen with the commands `dv` (display view), or `dt` (display table) since the view has not been logically modified, giving

| I | NZ | NE | ISLP | ILV | T | IT | LN | LL | AC | E(RYD) |
|---|---|---|---|---|---|---|---|---|---|---|
| 1 | 6 | 5 | 211 | 1 | T | 1 | 2 | 1 | | -1.78769E+00 |
| 2 | 6 | 5 | 211 | 2 | T | 1 | 3 | 1 | | -5.90082E-01 |
| 3 | 6 | 5 | 211 | 3 | T | 1 | 4 | 1 | | -3.10093E-01 |
| 4 | 6 | 5 | 211 | 4 | C | 2 | 0 | 0 | | -2.42561E-01 |
| 5 | 6 | 5 | 211 | 5 | T | 1 | 5 | 1 | | -1.89475E-01 |
| 6 | 6 | 5 | 211 | 6 | T | 2 | 3 | 0 | | -1.63384E-01 |
| 7 | 6 | 5 | 211 | 7 | T | 1 | 6 | 1 | | -1.24104E-01 |
| 8 | 6 | 5 | 211 | 8 | T | 1 | 7 | 1 | | -9.01943E-02 |
| 9 | 6 | 5 | 211 | 9 | T | 1 | 8 | 1 | | -6.82796E-02 |
| 10 | 6 | 5 | 211 | 10 | T | 1 | 9 | 1 | | -5.34454E-02 |
| 11 | 6 | 5 | 211 | 11 | T | 1 | 10 | 1 | | -4.29586E-02 |

The table lists for each state its term energy and configuration assignment. For a T-type state (see section 3), **IT** denotes the corresponding target-state index, and **LN** and **LL** represent, respectively, the active-electron $n$ and $l$ quantum numbers. For a C state, **IT** is an index in a list of configurations for the $(N+1)$-electron system. We note that there are two states that do not belong to the *np* series, namely **ILV=4** and **ILV=6**. The two records can be logically excluded from the view, thus creating a table, by typing the commands

```
ex (ilv=4
ex (ilv=6.
```

The command

```
dt nz ne islp ilv e qd iconf > prt
```

is then used to print a custom-made table containing the actual quantum defects and the electron configurations. It prints the table

| NZ | NE | ISLP | ILV | E(RYD) | QD | ICONF |
|---|---|---|---|---|---|---|
| 6 | 5 | 211 | 1 | -1.78769E+00 | 5.042E-01 | 2S2 2P |
| 6 | 5 | 211 | 2 | -5.90082E-01 | 3.964E-01 | 2S2 3P |
| 6 | 5 | 211 | 3 | -3.10093E-01 | 4.084E-01 | 2S2 4P |
| 6 | 5 | 211 | 5 | -1.89475E-01 | 4.053E-01 | 2S2 5P |
| 6 | 5 | 211 | 7 | -1.24104E-01 | 3.228E-01 | 2S2 6P |
| 6 | 5 | 211 | 8 | -9.01943E-02 | 3.405E-01 | 2S2 7P |
| 6 | 5 | 211 | 9 | -6.82796E-02 | 3.461E-01 | 2S2 8P |
| 6 | 5 | 211 | 10 | -5.34454E-02 | 3.488E-01 | 2S2 9P |
| 6 | 5 | 211 | 11 | -4.29586E-02 | 3.505E-01 | 2S2 10P |

Finally, a plot of quantum defects (qd) vs energy (e) for states contained in the table is displayed with the graphic command

```
pt e qd (fm=m,nl hd='QUANTUM DEFECTS FOR C II NP 2PO'
```

which displays the following headed graph with marked points but no joining lines.

## Acknowledgements

I am grateful to all participants of the OP for enjoyable and fruitful team work, specially to Professor M.J. Seaton as project leader. My participation in the OP has been fully supported by IBM Venezuela, to which I am indebted. I thank Dr. Donald C. Morton for permission to quote from his f-value compilation prior to publication.

## References

1. D.G. Hummer and D. Mihalas: *Astrophys. J.* **331**, 794 (1988)
2. D. Mihalas, W. Däppen and D.G. Hummer: *Astrophys. J.* **331**, 815 (1988)
3. W. Däppen, D. Mihalas, D.G. Hummer and B. Weibel-Mihalas: *Astrophys. J.* **332**, 261 (1988)
4. D. Mihalas, D.G. Hummer, B. Weibel-Mihalas and W. Däppen: *Astrophys. J.* **350**, 300 (1990)
5. M.J. Seaton: *J. Phys. B* **20**, 6363 (1987) (ADOC-I)
6. K.A. Berrington, P.G. Burke, K. Butler, M.J. Seaton, P.J. Storey, K.T. Taylor and Yu Yan: *J. Phys. B* **20**, 6379 (1987) (ADOC-II)
7. Yu Yan, K.T. Taylor and M.J. Seaton: *J. Phys. B* **20**, 6399 (1987) (ADOC-III)
8. Yu Yan and M.J. Seaton: *J. Phys. B* **20**, 6409 (1987) (ADOC-IV)
9. M.J. Seaton: *J. Phys. B* **20**, 6431 (1987) (ADOC-V)
10. J.F. Thornbury and A. Hibbert: *J. Phys. B* **20**, 6447 (1987) (ADOC-VI)
11. J.A. Fernley, K.T. Taylor and M.J. Seaton: *J. Phys. B* **20**, 6457 (1987) (ADOC-VII)
12. M.J. Seaton: *J. Phys. B* **21**, 3033 (1988) (ADOC-VIII)
13. G. Peach, H.E. Saraph and M.J. Seaton: *J. Phys. B* **21**, 3669 (1988) (ADOC-IX)
14. D. Luo, A.K. Pradhan, H.E. Saraph, P.J. Storey and Yu Yan: *J. Phys. B* **22**, 389 (1989) (ADOC-X)
15. D. Luo and A.K. Pradhan: *J. Phys. B* **22**, 3377 (1989) (ADOC-XI)
16. M.J. Seaton: *J. Phys. B* **22**, 3603 (1989) (ADOC-XII)
17. M.J. Seaton: *J. Phys. B* **23**, 3255 (1990) (ADOC-XIII)
18. J.A. Tully, M.J. Seaton and K.A. Berrington: *J. Phys. B* **23**, 3811 (1990) (ADOC-XIV)
19. A.K. Pradhan: *Physica Scripta* **35**, 840 (1987)
20. C. Mendoza: *Rev. Mex. Astron. Astrofis.* **21**, 613 (1990)
21. M.J. Seaton, C.J. Zeippen, J.A. Tully, A.K. Pradhan, C. Mendoza, A. Hibbert and K.A. Berrington: *Rev. Mex. Astron. Astrophys.*, in press
22. N.R. Simon: *Astrophys. J.* **260**, L87 (1982)
23. P.G. Burke and M.J. Seaton: *Meth. Comput. Phys.* **10**, 1 (1971)
24. W. Eissner, M. Jones and H. Nussbaumer: *Comput. Phys. Commun.*

8, 270 (1974)
25. A. Hibbert: *Comput. Phys. Commun.* **9**, 141 (1975)
26. P.G. Burke, A. Hibbert and W.D. Robb: *J. Phys. B* **4**, 153 (1971)
27. M.J. Seaton: in *Atomic Spectra and Oscillator Strengths for Astrophysics and Fusion Research*, J.E. Hansen (ed.), North-Holland, Amsterdam, 120 (1990).
28. K. Butler, C. Mendoza and C.J. Zeippen: in *Atomic Spectra and Oscillator Strengths for Astrophysics and Fusion Research*, J.E. Hansen (ed.), North-Holland, Amsterdam, 124 (1990).
29. C. Mendoza: in *Atomic Spectra and Oscillator Strengths for Astrophysics and Fusion Research*, J.E. Hansen (ed.), North-Holland, Amsterdam, 126 (1990).
30. H.E. Saraph and P.J. Storey: in *Atomic Spectra and Oscillator Strengths for Astrophysics and Fusion Research*, J.E. Hansen (ed.), North-Holland, Amsterdam, 128 (1990).
31. K. Butler and C.J. Zeippen: *Astron. Astrophys.* **234**, 569 (1990)
32. N. Allard, M.C. Artru, T. Lanz and M. Le Dourneuf: *Astron. Astrophys. Suppl. Ser.* **84**, 563 (1990)
33. K. Butler, C. Mendoza and C.J. Zeippen: in *Third Atomic Data Workshop*, C.J. Zeippen and M. Le Dourneuf (eds.), *J. Physique IV*, **1**, C1-135 (1991)
34. K. Butler and C.J. Zeippen: in *Third Atomic Data Workshop*, C.J. Zeippen and M. Le Dourneuf (eds.), *J. Physique IV*, **1**, C1-141 (1991)
35. A.K. Pradhan: in *Third Atomic Data Workshop*, C.J. Zeippen and M. Le Dourneuf (eds.), *J. Physique IV*, **1**, C1-153 (1991)
36. H.E. Saraph: in *Third Atomic Data Workshop*, C.J. Zeippen and M. Le Dourneuf (eds.), *J. Physique IV*, **1**, C1-157 (1991)
37. J.A. Tully, M.J. Seaton and K.A. Berrington: in *Third Atomic Data Workshop*, C.J. Zeippen and M. Le Dourneuf (eds.), *J. Physique IV*, **1**, C1-169 (1991)
38. W.C. Martin: *J. Phys. Chem. Ref. Data* **2**, 257 (1973)
39. A. Kono and S. Hattori: *Phys. Rev. A* **29**, 2981 (1984)
40. A. Kono and S. Hattori: *Phys. Rev. A* **31**, 1199 (1985)
41. A. Kono and S. Hattori: *Phys. Rev. A* **34**, 1727 (1986)
42. J.S. Sims and W.C. Martin: *Phys. Rev. A* **37**, 2259 (1988)
43. W.C. Martin: *Phys. Rev. A* **29**, 1883 (1984)
44. C.J. Sansonetti and W.C. Martin: *Phys. Rev. A* **29**, 159 (1984)
45. W.C. Martin: *Phys. Rev. A* **36**, 3575 (1987)
46. C.E. Moore: *Selected Tables of Atomic Spectra*, NSRDS-NBS (U.S.) 3, Sec. 3 (1970)
47. L. Johansson: *Arkiv Fysik* **31**, 201 (1965)
48. C.E. Moore: *Selected Tables of Atomic Spectra*, NSRDS-NBS (U.S.) 3, Sec. 4 (1971)
49. C.E. Moore: *Selected Tables of Atomic Spectra*, NSRDS-NBS (U.S.) 3, Sec. 5 (1975)

50. K.B.S. Eriksson: *Physica Scripta* **9**, 151 (1974)
51. K.B.S. Eriksson: *Physica Scripta* **28**, 593 (1983)
52. K.B.S. Eriksson: *Physica Scripta* **34** 211 (1986)
53. A.M. Malvezzi: *Physica Scripta* **27**, 413 (1983)
54. C.E. Moore: *Selected Tables of Atomic Spectra*, NSRDS-NBS (U.S.) 3, Sec. 7 (1976)
55. C.E. Moore: *Selected Tables of Atomic Spectra*, NSRDS-NBS (U.S.) 3, Sec. 9 (1980)
56. C.E. Moore: *Selected Tables of Atomic Spectra*, NSRDS-NBS (U.S.) 3, Sec. 10 (1983)
57. C.E. Moore: *Selected Tables of Atomic Spectra*, NSRDS-NBS (U.S.) 3, Sec. 11 (1985)
58. I. Wenåker: *Physica Scripta* **42**, 667 (1990)
59. W.C. Martin and R. Zalubas : *J. Phys. Chem. Ref. Data* **12**, 323 (1983)
60. A. Hibbert: *Rep. Prog. Phys.* **38**, 1217 (1975)
61. C. Froese Fischer and J.E. Hansen: *J. Phys. B* **18**, 4031 (1985)
62. C. Mendoza: *PhD Thesis*, University of London, UK (1979)
63. W.H. Parkinson, E.M. Reeves and F.S. Tomkins: *J. Phys. B* **9**, 157 (1976)
64. W.L. Wiese, M.W. Smith and B.M. Miles: *Atomic Transition Probabilities*, NSRDS-NBS (U.S.) 22, (1969)
65. P. Scott, A.E. Kingston and A. Hibbert: *J. Phys. B* **16**, 3945 (1983)
66. R.W. Ditchburn and R.D. Hudson: *Proc. R. Soc. A* **256**, 53 (1960)
67. G.H. Newsom: *Proc. Phys. Soc* **87**, 975 (1966)
68. V.L. Carter, R.D. Hudson and L.L. Breig: *Phys. Rev. A* **4**, 821 (1971)
69. G.A. Victor, R.F. Stewart and C. Laughlin: *Astrophys. J. Suppl. Ser.* **31**, 237 (1976)
70. C. Froese Fischer: *Can. J. Phys.* **53**, 184 (1975)
71. C. Froese Fischer: *Can. J. Phys.* **53**, 338 (1975)
72. C. Froese Fischer: *J. O. S. A.* **69**, 118 (1979)
73. C. Froese Fischer and M. Godefroid: *Nucl. Instr. Meth.* **202**, 307 (1982)
74. K.L. Baluja and A. Hibbert: *J. Phys. B* **13**, L327 (1980)
75. K.L. Baluja and A. Hibbert: *Nucl. Instr. Meth.* **B9**, 477 (1985)
76. S.S. Tayal and A. Hibbert: *J. Phys. B* **17**, 3835 (1984)
77. S.S. Tayal: *J. Phys. B* **19**, 3421 (1986)
78. R. Moccia and P. Spizzo: *J. Phys. B* **21**, 1133 (1988)
79. D.C. Morton: *Astrophys. J. Suppl. Ser.*, in press
80. R.L. Kurucz and E. Peytremann: *Smithsonian Astrophysical Obs. Special Report* No. 362 (1975)
81. R.L. Kurucz: private communication to D.C. Morton.
82. D. Luo, A.K. Pradhan and J.M. Shull: *Astrophys. J.* **335**, 498 (1988)
83. P.L. Dufton, A. Hibbert, A.E. Kingston and J.A. Tully: *M. N. R. A. S.* **202**, 145 (1983)
84. J.M. Shull, T.P Snow and D.G. York: *Astrophys. J.* **246**, 549 (1981)
85. I. Martinson: *Rep. Prog. Phys.* **52**, 157 (1989)

86. Y. Baudinet-Robinet, H.P. Garnir, P.D. Dumont and J. Résimont: *Phys. Rev. A* **42**, 1080 (1990)
87. Y. Baudinet-Robinet, P.D. Dumont and H.P Garnir: *Phys. Rev. A* **43**, 4022 (1991)
88. C.E. Theodosiou: *Phys. Rev. A* **30**, 2910 (1984)
89. N. Reistad, R. Hutton, A.E. Nilsson, I. Martinson and S. Mannervik: *Physica Scripta* **34**, 151 (1986)
90. D.J. Bradley, C.H. Dugan, P. Ewart and A.F. Purdie: *Phys. Rev. A* **13**, 1416 (1976)
91. A. Dasgupta and A.K. Bhatia: *Phys. Rev. A* **31**, 759 (1985)
92. C. Mendoza and C.J. Zeippen: *Astron. Astrophys.* **179**, 346 (1987)
93. R.E. Bonanno, C.W. Clark and T.B. Lucatorto: *Phys. Rev. A* **34**, 2082 (1986)
94. J.B. West and G.V. Marr: *Proc. R. Soc. Lond. A* **349**, 397 (1976)
95. A.M. Cantú, M. Mazzoni, M. Pettini and G.P. Tozzi: *Phys. Rev. A* **23**, 1223 (1981)
96. J.A.R. Samson and G.C. Angel: *Phys. Rev. A* **42**, 1307 (1990)
97. G.C. Angel and J.A.R. Samson: *Phys. Rev. A* **38**, 5578 (1988)
98. R.D. Hudson and V.L. Carter: *J. O. S. A.* **57**, 651 (1967)
99. J.L. Kohl and W.H. Parkinson: *Astrophys. J.* **184**, 641 (1973)
100. R.A. Roig: *J. Phys. B* **8**, 2939 (1975)
101. R.F. Reilman and S.T. Manson: *Astrophys. J. Suppl. Ser.* **40**, 815 (1979)
102. W. Cunto and C. Mendoza: *TOPBASE User's Manual*, Report CSC-02-91, IBM Caracas Scientific Center, Venezuela (1991)
103. W. Cunto and C. Mendoza: *Rev. Mex. Astron. Astrofis.*, in press

# 8

# Sources of Atomic Spectroscopic Data for Astrophysics

W. C. Martin[1]

## I. Introduction

This article is mainly a bibliographic review of compilations of data on atomic energy levels and wavelengths (Part II), transition probabilities (Part III), and spectral line shapes and shifts (Part IV). The selection of references is based on the needs of astronomers, and guidance to the original literature on laboratory data is highly selective or, for some types of data, limited to the citation of appropriate bibliographies and reviews. Several of the larger collections of computer-readable atomic data are described briefly in Part V. Although the length of this article does not permit a review of the published sources of data on photoionization and atomic collisions, information on several databases covering these subjects is included in Part V.

## II. Published Data on Energy Levels and Wavelengths

Sections A through F below cover the elements in order of increasing atomic weight, the references being mainly limited to compilations of critically evaluated data now included in the NIST Database (see Part V) and to a selection of references to more recent laboratory research of astrophysical interest. Although no data from Moore's older compilations *Atomic Energy Levels* [1949–1958], *A Multiplet Table of Astrophysical Interest* [1945], and *An Ultraviolet Multiplet Table* [1950–1962] have been included

---

[1] National Institute of Standards and Technology, Gaithersburg, Maryland 20899, U.S.A.

thus far, these classic tables have by no means been entirely superseded. The NIST compilations of energy level data carried out during the past fifteen years or so have together covered a considerably larger part of the periodic table than has yet been covered by corresponding new wavelengths compilations based on the energy levels evaluations. Fairly complete references for laboratory wavelength data are, however, given in the published energy level compilations.

With regard to the remaining sections of Part II, section G covers ionization energies. Section H is organized according to isoelectronic sequence and covers the hydrogen through sodium sequences. Additional tables of wavelengths and wavelength standards are cited in section I, and section J gives references to bibliographic publications. The triennial IAU reports cited in section J list recent references for original laboratory research on energy levels and wavelengths of strong astrophysical interest for the elements up through nickel (Z=28). Only a few such references are given here (sections A, B, and C).

## *A. Hydrogen through neon (Z=1−10)*

Moore's *Selected Tables of Atomic Spectra* [1965−1985] include compilations of energy levels and multiplet tables for the spectra H I, C I−VI, N I−VII, O I, O III−VIII, and Si I−IV. New observations and an extension of the O II analysis have been completed [Eriksson 1987, Pettersson and Wenåker 1990, Wenåker 1990], and Eriksson [1986] gives new wavelengths for N I in the vacuum ultraviolet.

Accurate values and series formulas for all single-excitation energy levels of He I are available [Martin 1987], and a compilation of wavelengths and levels for B I has been published [Odintzova and Striganov 1979]. Persson *et al.* [1991] recently completed a large extension of the observations and analysis of Ne III.

Highly accurate values for ground-term level separations are important for some spectra in connection with the identification and analysis of far-infrared lines observed, for example, in interstellar sources. Fairly recent determinations of such separations have been published for C I [Cooksy *et al.* 1986c], C II [Cooksy *et al.* 1986a], N II [Cooksy *et al.* 1986b], and O I [Zink *et al.* 1991].

## *B. Sodium through argon (Z=11−18)*

During the past twelve years new energy-level compilations have been published for all spectra of sodium (Na I−XI), magnesium (Mg I−XII), aluminum (Al I–XIII), silicon (Si I–XIV) [Martin and Zalubas 1981, 1980, 1979, 1983, resp.], phosphorus (P I–XV) and sulfur (S I–XVI) [Martin *et al.* 1985, 1990]. More recently we have begun compiling wavelengths with energy-level classifications for the spectra of this group of elements: the data for magnesium (Mg I–XII) and aluminum (Al I–XIII) have been published [Kaufman and Martin 1991a, 1991b], the compilations for sulfur are almost complete, and work on the sodium and silicon compilations is underway. All wavelength regions from x-rays to the far infrared are covered. In order to produce more accurate and complete wavelength data, we are updating and extending our energy-level compilations for many of the spectra. In addition to the observed wavelengths we give wavelengths calculated from the optimized energy levels; the calculated wavelengths are in many cases more accurate than the observed values. We also use the levels to give reliable predicted wavelengths for many lines (allowed and forbidden) that have not been observed in the laboratory, but which occur or are predicted to occur in astronomical spectra.

### *C. Potassium through nickel (Z=19–28)*

Sugar and Corliss [1985] have compiled the energy level data for all spectra of these elements. Wavelength compilations based on these critically reviewed data have been carried out for the spectra of scandium (Sc I–XXI) [Kaufman and Sugar 1988] and for the higher spectra of vanadium (V VI–XXIII), chromium (Cr V–XXIV), iron (Fe VIII–XXVI), cobalt (Co VIII–XXVII), and nickel (Ni IX–XXVIII) [Shirai *et al.* 1992b, 1992c, 1990, 1992a, 1987a, resp.]. Johansson and Cowley [1988] reviewed the available data for the first three spectra of the iron-group elements scandium through nickel and assessed the completeness of the line lists and analyses in view of astrophysical needs. A similar review of the analyses of the third and fourth spectra of vanadium through nickel was published by Cowley and Frey [1988].

Some recent and ongoing research on the first two spectra of iron-group elements is noted in a recent review of high-resolution atomic spectroscopy in the infrared region [Johansson 1991]. These investigations include an extension of the Ti I analysis [Forsberg *et al.* 1986, Forsberg 1991], and ongoing work on the analysis of Ti II, Cr I, II, Fe I, II (see below), Co II, and Ni I. Accurate laboratory measurements have been published for lines of Cr I, II and Fe I, II [Biémont *et al.* 1985b, 1985a], and Ni I, II [Biémont *et al.* 1986, Litzén 1990] that have been identified in solar spectra for the infrared range 9000–1800 $cm^{-1}$.

No complete accounts of recent work on the astrophysically important spectra Fe I and Fe II have yet been published, but a few references will be mentioned here and in following sections on reference wavelengths and transition probabilities. A paper by Brown *et al.* [1988] on Fe I absorption in the 1500–3215 Å region includes some 3000 lines and many new levels. All other recent wavelength measurements for Fe I and Fe II cited here have been made with high-accuracy Fourier-transform-spectrometric (FTS) techniques. Johansson and Learner [1990] and Johansson *et al.* [1991a] have classified additional Fe I infrared lines and derived new levels using accurate FTS data, including the measurements of Biémont *et al.* [1985a]. Adam *et al.* [1987] gave newly classified Fe II lines (1306–2804 Å) and new energy levels, Johansson and Baschek [1988] published additional new Fe II levels, and Rosberg and Johansson [1992] have recently further extended the Fe II analysis with new line classifications in the region near 10,000 $cm^{-1}$. Some 20 newly classified Fe II lines in the range 8067–8847 Å have been tabulated [Johansson *et al.* 1991b].

A monograph by Iglesias *et al.* [1988] includes all data from the analysis of V II, with some 2800 classified lines, and the results of an extension of the Ni III analysis have been described and tabulated by Garcia-Riquelme and Rico [1991]. Ekberg *et al.* [1990] have classified more than 500 lines of Ca IV in the range 295–2830 Å. Sugar *et al.* [1992] recently measured the wavelengths of prominent lines of Ni X to Ni XXVI (83–320 Å) with significantly improved accuracy.

## D. *Copper through barium (Z=29–56)*

New compilations of energy levels have been published for all known spectra of copper, krypton, and molybdenum [Sugar and Musgrove 1990, 1991, 1988, resp.], and a similar compilation for the germanium spectra is forthcoming. Wavelength compilations based on these data have been published for the higher spectra of copper (Cu X–XXIX) [Shirai *et al.* 1991] and molybdenum (Mo VI–XLII) [Shirai *et al.* 1987b], and similar compilations for the higher ions of krypton and germanium are underway. Only a few of the more recent references for spectra of other elements can be mentioned here. The analyses of Ga II [Isberg and Litzén 1985] and Ga III [Isberg and Litzén 1986, Ryabtsev and Wyart 1987] have been extended. An extensive new analysis of Y II has been completed [Nilsson *et al.* 1991], and research on Zr II is underway. A revised analysis of Xe II was published by Hansen and Persson [1987].

## E. *Lanthanum through lutecium (Z=57−71)*

A compilation of energy level data for the known spectra of these rare-earth elements was published in 1978 [Martin *et al.*]. Spectra for which extensions or revisions of the analyses have been published more recently include Pr I [Ginibre 1981], Pr II [Ginibre 1989a, 1989b, 1990], Nd II [Blaise *et al.* 1984], Eu I [Wyart 1985], Yb I [Meggers and Tech 1978, Wyart and Camus 1979, Aymar *et al.* 1984], Yb II [Wyart and Camus 1979, Sugar and Kaufman 1979, Wilson 1990], and Lu III [Sugar and Kaufman 1979].

## F. *Hafnium through einsteinium (Z=72−99)*

Vol. III of Moore's *Atomic Energy Levels* includes the data for hafnium through actinium (Z=72−89) as available in 1958. Raassen *et al.* [1990] have recently reviewed the availability of analyses of these elements up through bismuth (Z=83). Several new investigations of the platinum spectrum have yielded much more accurate data for both Pt I [Engleman 1985, Reader *et al.* 1990a, Blaise *et al.* 1992] and Pt II [Reader *et al.* 1988, 1990a, Engleman 1989, Sansonetti *et al.* 1992], and the Pt I and Pt II analyses have been significantly extended [Blaise *et al.* 1992, Blaise and Wyart 1992b]. The wavelength data given in some of these references are described below (see *Wavelength standards*). Other examples of recent research are new analyses of W III [Iglesias *et al.* 1989], Au II, Hg III, Tl IV, Pb V, Bi VI [Wyart *et al.* 1992], Hg IV [Joshi *et al.* 1989, van der Valk *et al.* 1990] and studies of the Bi I absorption spectrum in the vacuum ultraviolet [Mazzoni *et al.* 1987, Mathews *et al.* 1989].

Data for the actinide elements actinium through einsteinium (Z=89−99) and for francium (Z=87) have been compiled by Blaise and Wyart [1992a], who give the experimental energy levels for all known spectra of these elements. Hyperfine structure and isotope-shift data are included, and also wavelengths and classifications for the stronger lines of the actinide spectra.

## G. *Ionization energies*

All of the NIST compilations of energy level and/or wavelength data include the ionization energy for each spectrum; in general, the most recent publication should be consulted. Moore's [1970a] compilation of ionization energies included the ground-term levels for all spectra for which the data were then available.

## H. *Compilations of data for isoelectronic sequences*

Analysis of results along isoelectronic sequences is one of the most effective methods for evaluating atomic data. In carrying out the compilations by element described above we have taken into account data from isoelectronic studies, including semiempirical treatments of the lithium through fluorine sequences and the sodium sequence. Some isoelectronic data sources for the hydrogen through sodium sequences are cited below. These references are especially useful for isoelectronic studies, of course, and for data on spectra not yet included in more comprehensive compilations. Many additional references can be found in the review literature; Martinson [1989], for example, discusses research on the spectra of highly ionized atoms by organizing many of the results according to isoelectronic sequences.

### 1. Hydrogen sequence

In general the most accurate data are from theory. Johnson and Soff [1985] give the 1s, 2s, and 2p levels for the elements $Z=1-110$. Mohr [1983] gives the equivalent data for $Z=10-40$. The older calculations of Erickson [1977] are sufficiently accurate for most purposes: he gives all levels having $n\leq 11$ for the elements $Z=1-15$, those having $n\leq 5$ for $Z=16-39$, and those having $n\leq 3$ for $Z=40-105$.

### 2. Helium sequence

The accuracy of theoretically calculated levels is comparable to or better than the accuracy of the available experimental values for most of the heavier members of this sequence. Drake [1988] gives calculated levels for the $1s^2$, 1s2s, and 1s2p configurations of all helium-like spectra for $Z=2-100$. Vainshtein and Safronova [1985] tabulate calculated levels of the $1sn\ell$ configurations ($\ell$=s,p,d) through $n=5$ for $Z=6-42$. Series formulas for the $1sn\ell$ levels ($\ell$=s,p,d,f) have been given for $Z=11-18$ [Martin 1981], and the $1sn\ell$ term values for $\ell\geq 4$ are equal to the appropriate hydrogenic values to a good approximation for these elements.

Transitions of the type $1sn'\ell'-n\ell n'\ell''$ ($n,n'\geq 2$) from doubly-excited configurations in helium-like ions give rise to satellite lines near the 1s−np resonance lines of the corresponding hydrogen-like ions. Vainshtein and Safronova [1978, 1980] gave calculated wavelengths for such satellites due to $n\ell n'\ell''$ configurations having $n,n'\leq 3$ in all helium-like ions from carbon through arsenic ($Z=6-33$). The results of these calculations as used in some of the more recent NIST compilations have been adjusted to include QED energies [see, e.g., Martin *et al.* 1990, Kaufman and Martin 1991a, 1991b].

## 3. Lithium sequence

Edlén's semiempirical formulas for this sequence cover the entire $1s^2n\ell$ energy structure [1979, 1983a]. The data he tabulated for Z=3−28 included experimental wavelengths available at the time. Vainshtein and Safronova [1985] give calculated levels for the $1s^2n\ell$ configurations up to n=5 with $\ell$=s,p,d for the elements Z=6−42. Kim *et al.* [1991] compare new theoretical calculations of the 2s−2p separations with available experimental values for Z=6−92 and give accurate predicted values for the ions up through Z=60.

Transitions of the type $1s^2n'\ell'-1sn\ell n'\ell''$ (n,n'≥2) from K-shell excited configurations in lithium-like ions give rise to satellite lines near the $1s^2-1sn\ell$ lines of the corresponding helium-like ions. (The astrophysically most important satellites of this type belong to $1s^2n'\ell'-1s2pn'\ell'$ arrays and lie on the long-wavelength side of the $1s^2\ {}^1S_0^{\circ}-1s2p\ {}^1P_1^{\circ}$ resonance line.) Calculated wavelengths for such satellite transitions in spectra of astrophysical interest have been tabulated by, for example, Vainshtein and Safronova [1978, 1980], Hata and Grant [1984], Chen [1986], and Nilsen [1988]. Karim and Bhalla [1990] give such calculated wavelengths for higher n' values (n'=5−8) in helium-like chromium and nickel, and include a useful set of references to other calculations. Bitter *et al.* [1991] cite several recent papers comparing theory and experiment for satellites near the Ni XXVII $1s^2-1s2p\ {}^1P_1^{\circ}$ resonance line.

## 4. Beryllium through fluorine sequences

Edlén's [1983−1985] compilations and regularizations of data for these sequences include the energy levels and wavelengths for transitions involving the $2s^22p^N$, $2s2p^{N+1}$ and $2p^{N+2}$ configurations. The references sorted according to sequence are: [1983a] and [1985b] for the beryllium sequence, [1983b] for the boron sequence, [1985a] for the carbon sequence, [1984] for the nitrogen sequence, and [1983a] for the oxygen and fluorine sequences. The data extend to krypton (Z=36) and in some cases to molybdenum (Z=42). The theoretical results obtained by Kim *et al.* [1988] for the beryllium-sequence $2s^2-2s3p$ transitions include accurate wavelengths for Mg IX through Kr XXXIII.

Transitions of the type $1s^22s^M2p^N-1s2s^M2p^{N+1}$ in beryllium- through fluorine-like ions can contribute additional satellites and blended features on the long-wavelength side of the $1s^2-1s2p$ resonance lines of the corresponding helium-like ions. Calculated wavelengths for such transitions in beryllium-like ions (M+N=2) are given, for example, by Safronova and Lisina [1979] and by Hata and Grant [1984]. The most detailed comparisons of observations and calculations for K-shell transitions of the above types have been given in papers with results for only one or two elements. Seely *et al.* [1986], for example, tabulate such comparisons for Fe XVIII−XXIV as measured in solar flare spectra and include references to other observations and calculations of these and similar transitions.

5. Neon sequence

Fawcett's [1984] paper giving calculated oscillator strengths for Al IV through Ar IX (Z=13−18) includes a compilation of the experimental $2p^5 3\ell$ levels for these spectra. Buchet *et al.* [1987] include a compilation of the $2p^5 3\ell$ levels for Ar IX through Cu XX (Z=18−29). Wavelengths and energy levels for lines belonging to the $2p^5 3\ell - 2p^5 4\ell'$ transition arrays have been given for Ca XI through Mn XVI (Z=20−25) by Jupén *et al.* [1987].

6. Region below 30 Å, lithium through neon sequences

A book by Boiko *et al.* [1988] includes compilations of wavelengths in the X-ray region 1−26 Å for the lithium through neon sequences; the data for the neon sequence, for example, extend up to neodymium (Z=60).

7. Sodium sequence

Edlén [1978] gave data and formulas for the 3s, 3p, 3d, and $n\ell$ ($\ell \geq 3$) levels and transitions for the elements sulfur through molybdenum (Z=16−42). His data for some of these ions have been superseded, and accurate measurements and predictions are now available for heavier elements: see, e.g., Kim *et al.* [1991] for the 3s−3p separations, and Reader *et al.* [1987, 1990b] and Seely *et al.* [1991] for the levels and wavelengths of transitions involving the 3d and higher configurations.

## *I. Some other published tables of wavelengths and wavelength standards*

Several of the more extensive compilations of wavelength data are noted here. The reader should also consult the references cited in Part III (Transition Probabilities); several of those publications are also useful for line identifications, and some include the most accurate available wavelengths for the tabulated lines.

1. Wavelength tables having broad wavelength coverage

The tables edited by Reader *et al.* [1980] have almost 47,000 lines arranged by spectrum and also in a finding list. The stronger lines observed for the first five spectra of all elements are listed, the coverage extending over all wavelength regions. The first part of these tables (lines arranged by spectrum) are included with some corrections and additions in the *CRC Handbook of Chemistry and Physics* [Reader and Corliss 1990]. Part 2 of the *Tables of Spectral Lines* by Zaidel' *et al.* [1970, 1977] has 38,000 lines of 98 elements arranged by element; the energies of the upper levels are included (to the nearest 0.01 eV) for the classified lines. The tables of Striganov and Odintsova [1982] include all spectra of hydrogen through neon, the first through the sixth spectra of sodium through chlorine, and Ar I through Ar X. A total of 30,490 lines are listed, by spectrum, with level classifications and excitation potentials. These tables supersede the tables of Striganov and Sventitskii [1968] for the spectra mentioned, but the older publication is more widely available and includes the first few spectra of several elements heavier than argon.

2. Wavelengths below 3000 Å

R. L. Kelly's [1987] compilation of atomic lines below 2000 Å covers all ionization stages of hydrogen through krypton and includes the observed lines (with energy-level classifications) sorted according to spectrum and a finding list. Kelly's 1979 compilation covers the spectra of these same elements over the wavelength range 2000–3200 Å, with some additional lines up to 3645 Å. The tables of Meggers *et al.* [1975], which include lines from seventy elements (see next section), also extend down to 2000 Å. C. E. Moore's [1950–1962] ultraviolet multiplet tables are well known to astronomers.

3. Wavelengths above 3000 Å

The *Tables of Spectral Line Intensities* of Meggers *et al.* [1975] include 39,000 lines of seventy elements in the region 2000–9000 Å. The lines were excited in an arc discharge and belong mainly to the first two spectra. Energy levels of classified lines are given to the nearest $cm^{-1}$, and the tables for several rare-earth spectra include lines and classifications that have not been published elsewhere. C. E. Moore's [1945] multiplet tables remain useful for many spectra in this region. Outred [1978] compiled wavelengths and wavenumbers for 8885 lines of 57 elements in the infrared region 10,000−2400 $cm^{-1}$, the lines belonging mainly to the first two spectra. Energy levels are given to the nearest $cm^{-1}$ for the classified lines, and a finding list is included.

## 4. Wavelength standards

A compilation of reference wavelengths by Kaufman and Edlén [1974] covers the range 15 to 25,000 Å, with most uncertainties ranging from 0.0001 to 0.002 Å. Their tables of wavelengths below 2000 Å have 2091 lines of 28 elements, and the tables of wavelengths above 2000 Å have 3341 lines of ten elements. The references cited by Kaufman and Edlén need not be repeated here.

Mount *et al.* [1977] first noted the usefulness of a platinum hollow-cathode lamp with neon carrier gas as a source for wavelength calibration. Reader *et al.* [1990a] give wavelengths for about 3000 lines from this source, with uncertainties of 0.002 Å or less over the range 1100−4000 Å. An atlas of the spectrum with wavelengths and intensities for about 6000 lines has also been prepared [Sansonetti *et al.* 1992]. Engleman's high-accuracy FTS measurements have been published for the ranges 2245−5223 Å for Pt II [Reader *et al.* 1988] and 2200−7219 Å for Pt I [Engleman 1985].

The number of accurately determined wavelengths of iron has been significantly increased since Kaufman and Edlén's [1974] compilation. Crosswhite [1975] listed some 4000 wavelengths for the spectrum of an iron hollow-cathode with neon carrier gas. The range is from 1934 to 9000 Å, with 2377 lines of Fe I and including hundreds of Fe I wavelengths accurate to a few ten-thousandths of an Å as calculated from the energy levels. Improved wavelengths for many Fe I lines can be obtained from the new level values of O'Brian *et al.* [1991]; these authors give 442 Fe I levels with errors mainly in the range 0.001 to 0.005 $cm^{-1}$ as determined from FTS measurements of a hollow-cathode spectrum. Nave *et al.* [1991] give newly determined wavelengths for 539 lines of Fe I and Fe II in the range 1830−3841 Å with estimated errors of 0.00008 to 0.00015 Å. Learner and Thorne [1988] measured the wavelengths of some 300 Fe I lines in the range 3824−5763 Å within an estimated error of 0.001 $cm^{-1}$ (±0.00015 to ±0.0003 Å). Johansson and Learner's [1990] table of Fe I lines in the infrared range 7320−4712 $cm^{-1}$ has about 360 lines, with estimated errors less than 0.003 $cm^{-1}$ for symmetric lines.

The spectrum of thorium emitted from a hollow-cathode or electrodeless discharge is a traditional source of reference wavelengths. Palmer and Engleman's *Atlas of the Thorium Spectrum* [1983] includes an extensive list of lines from 2777 to 13,500 Å with wavenumbers accurate to ±0.002 $cm^{-1}$ (±0.0005 Å at 5000 Å). Uranium excited in the above types of discharges also gives a very line-rich spectrum useful for wavelength calibration of high-resolution spectrometers. A similar atlas for uranium lists 4928 lines (3848–9084 Å) with wavenumbers accurate to ±0.003 $cm^{-1}$ [Palmer *et al.* 1980]. Sansonetti and Weber [1984] have measured a number of thorium and uranium lines (5750–6920 Å) with uncertainties of ±0.0004 $cm^{-1}$ for thorium and ±0.0003 $cm^{-1}$ for uranium. Their measurements show the wavenumber errors of the corresponding lines in the Los Alamos atlases to be well within the quoted uncertainty estimates.

### *J. Bibliographies*

The first three sections of Moore's *Bibliography on the Analyses of Optical Atomic Spectra* [1968, 1969] extend the bibliographies given in her three *Atomic Energy Levels* volumes to 1968, and a fourth section covers the literature for the rare-earth and actinide spectra to early 1969. The literature for all spectra from July 1968 through December 1983 is covered in a *Bibliography on Atomic Energy Levels and Spectra* and three supplements [Hagan and Martin 1972, Hagan 1977, Zalubas and Albright 1980, Musgrove and Zalubas 1985]. Bibliographic reports covering the data of particular interest for astronomy as published up through mid-1990 are given in the triennial IAU Transactions [McNally 1991]. More complete coverage of the literature for elements heavier than nickel can be found in a series of bulletins on atomic and molecular data for fusion that now appear twice yearly [Smith 1991]; these have a bibliographic section covering energy levels and wavelengths for selected elements.

Heilig [1977–1992] has published bibliographies on optical isotope shifts covering the literature from 1918 to May 1991.

## III. Published Data on Transition Probabilities

The references cited in this section are mainly limited to the more extensive compilations of evaluated data and to some other available sources of data for many elements. Only a few references for data on particular spectra are cited as examples of recent research.

### *A. Compilations of evaluated data*

Two volumes of critically evaluated data on transition probabilities compiled by Martin *et al.* [1988] and Fuhr *et al.* [1988] include 8800 lines for the elements scandium through manganese (Z=21−25) and 9500 lines for iron through nickel (Z=26−28). The lines are tabulated by spectrum and are also separated into allowed and forbidden transitions.

Another recent compilation gives evaluated transition probability data for about 8300 selected lines of all elements for which reliable data are available [Fuhr and Wiese 1990]. These data are mainly for the stronger lines of the first two spectra but include some prominent lines for higher ionization stages of important elements. The selected data for the elements hydrogen through calcium include some material from two earlier and generally more extensive compilations [Wiese *et al.* 1966, 1969], but also comprise more recent results from the literature.

Morton's [1991] recent compilation of data for 2031 resonance lines covers all spectra of hydrogen through germanium (Z=1−32) having allowed absorption transitions originating on ground-term levels with wavelengths above the Lyman limit at 911.75 Å. Carefully selected transition-probability values are included for most of the lines, radiation damping constants are given for many of the stronger ones, and the wavelengths are the most accurate available. Morton plans to compile similar data for the range from the He II limit at 227.8 Å up to 911.75 Å.

## B. *Opacity Project*

Calculations carried out under the Opacity Project [Seaton 1987] have given new transition probability data for spectra of the elements hydrogen through neon and for iron. These results are described elsewhere in this volume [Mendoza 1992].

## C. *"Solar" oscillator strengths*

Thévenin [1989, 1990] has most recently used solar-spectrum data to derive an extensive set of oscillator strengths. He tabulates values for 4597 lines of 35 elements in the range 4006−5684 Å and for more than 1500 lines in the range 5684−7950 Å.

## D. *Forbidden lines*

Kaufman and Sugar [1986] have compiled tables of lines due to magnetic-dipole transitions in spectra of the elements beryllium through molybdenum (Z=4−42). The lines arise from observed or predicted transitions between levels of ground configurations of the types $ns^2np^N$ (n=2 and 3). The data include calculated values for the transition probabilities of 1660 lines in the range 100 Å to 25.9 mm. The wavelengths given for these lines were in general the most accurate available values.

Biémont and Hansen [1986] give calculated transition probabilities for both magnetic-dipole and electric-quadrupole lines within the $p^4$ ground configurations of the sulfur and selenium isoelectronic sequences up to molybdenum and silver, respectively. References for similar calculations for other $p^N$ ground configurations are included in their paper.

## E. *Two older publications of data for many elements*

Corliss and Bozman [1962] published transition probabilities for some 25,000 lines from 112 spectra (mainly first and second spectra) of seventy elements. These data were derived from the measured relative intensities of lines emitted from an arc source (2000–9000 Å) [Meggers *et al.* 1975]. Unfortunately, many of the resulting transition probabilities for important spectra such as those of the iron-group elements were later shown to be in error by factors of ten or more [see, e.g., Bridges and Kornblith 1974 and references therein].

The original tables of Kurucz and Peytremann [1975] give gf values for 265,587 classified lines belonging mainly to the first five or six spectra of the elements helium through nickel. The wavelength range is 52 Å to 10 $\mu$m. Except for about 10,000 values taken from the literature, the gf values were calculated semiempirically. The tables include earlier published gf values for elements heavier than nickel. Kurucz has more recently carried out large-scale extensions and revisions of the data for many spectra (see below).

## F. *Isoelectronic sequences*

The literature on theoretically calculated transition probabilities for isoelectronic sequences is quite extensive; the number of papers giving experimental results for sequences is much smaller. A considerable selection of the most accurate of these data has been included in the compilations cited above. The references for isoelectronic sequences will not be reviewed here, but the reader will find guidance to the literature in the compilations and in bibliographies and reviews cited in section H.

## G. *Examples of recent research: N I, Si I, Fe I,II*

Several references on N I can be cited as examples of recent experimental and theoretical research on accurate transition probabilities for astrophysically important 2p-shell spectra. Goldbach *et al.* [1986, 1988] measured oscillator strengths for lines of ten N I vacuum-ultraviolet multiplets using a wall-stabilized arc source, and Zhu *et al.* [1989] obtained transition probabilities for 29 N I lines in the range 4137–8747 Å with the same method. Following the usual procedures in such work, both groups normalized their measurements of relative values to an absolute scale by using previous lifetime determinations: some methods used in relevant lifetime determinations were the beam-foil technique [e.g., Chang 1977], the phase-shift method [e.g., Brooks *et al.* 1977], and laser-induced fluorescence [e.g., Copeland *et al.* 1987]. Hibbert *et al.* [1991] and Suskin and Weiss [1991] have recently made accurate calculations of transition probabilities for astrophysically important N I lines. Their results for lines in common are generally in good agreement and also in most cases agree well with the experimental values cited above.

Smith *et al.* [1987] obtained experimental oscillator strengths for 108 Si I lines (1625–4103 Å) by combining hook-method absorption measurements with branching-ratio measurements. Improved values for the strengths of 36 of these lines have been derived by O'Brian and Lawler [1991] using their new lifetime determinations for 47 levels.

Recent research on the important Fe I and Fe II spectra illustrates a productive interplay between experimental measurements and theoretical calculations of transition probabilities. A paper by Biémont *et al.* [1991] includes new experimental lifetimes for twelve Fe II levels, theoretical gf values for 39 selected lines, and a useful set of references to other recent results on Fe II lifetimes and transition probabilities. Bard *et al.* [1991] derived gf values for 114 Fe I lines (2610–11,690 Å) using new emission measurements combined with lifetime data from the literature. By combining several techniques, O'Brian *et al.* [1991] have obtained experimental transition probabilities for 1814 lines of Fe I (2250–26,660 Å). These latter authors mention their frequent use of extensive unpublished calculations of Fe I transition probabilities by Kurucz [1991] as a guide in their measurements of this complex spectrum. Kurucz in turn continues to revise and extend his results for the iron-group spectra, including Fe I and especially Fe II [Kurucz 1981], using new data obtained in the ongoing extensions of the analyses of these spectra [Kurucz 1991, Johansson and Baschek 1988, Leckrone *et al.* 1990].

## *H. Bibliographies and reviews*

A bibliography on atomic transition probabilities by Fuhr *et al.* [1978b] and a supplement by Miller *et al.* [1980] cover the literature through March 1980. Bibliographic reports covering the most important data for astronomy are published triennially; the most recent such report [McNally 1991] updates the coverage of the literature to mid-1990. A series of bulletins appearing twice yearly has a bibliographic section covering transition probabilities for most elements of greatest astrophysical interest [Smith 1991].

Wiese and Fuhr [1990] have reviewed the overall availability of reliable transition probability data as of mid-1988.

The proceedings of a colloquium held in August 1989 include a number of review articles of interest, several of which can be mentioned here [Hansen 1990]. Blackwell's [1990] paper on the accuracy of furnace oscillator-strength measurements includes a useful set of references for data obtained by this method. Biémont [1990] gave an extensive list of references to both experimental and theoretical papers in his review of progress on oscillator-strength determinations for elements heavier than nickel ($Z>28$). A review by Martinson [1990] centered on experimental determinations of lifetimes and oscillator strengths for ions of the iron-group and rare-earth elements. Hibbert [1990] reviewed the calculation of accurate transition probabilities for weak lines, and Zeippen [1990] reviewed such calculations for forbidden lines.

# IV. Spectral Line Widths and Shifts

The literature on experimental determinations of Stark widths and shifts has been critically reviewed in five reports including tabulations of selected experimental Stark broadening parameters: the data for neutral atoms was covered by Konjević and Roberts [1976] and by Konjević *et al.* [1984a]; the data for ions was covered by Konjević and Wiese [1976] and by Konjević *et al.* [1984b]; and, most recently, Konjević and Wiese [1990] extended the coverage for both neutral atoms and ions through 1988.

Dimitrijević [1988] has applied a semiempirical method to obtain Stark broadening data for 127 multiplets of a number of astrophysically important third and fourth spectra. A semiclassical-perturbation formalism is being used in a new program for the calculation of such data for lines of multicharged ions of astrophysical interest; the first two of a series of planned papers have appeared, giving the results for C IV and Si IV lines [Dimitrijević *et al.* 1991a, 1991b].

Dimitrijević [1990] has reviewed the accuracy of line broadening data of various types, including Stark broadening of hydrogen and hydrogen-like lines. Mathys [1988] has emphasized that recent improvements in theories of ionic broadening should be applied to obtain more accurate data for hydrogen lines. Tables of Stark broadened profiles of the L$\alpha$, L$\beta$, and H$\alpha$

hydrogen lines in the presence of magnetic fields have been published [Mathys, 1985]. Allard and Kielkopf [1982] reviewed the effects of neutral nonresonant collisions on spectral lines.

In addition to references cited in the above reviews, much of the literature on line broadening data important for astrophysics has been included in bibliographies. A *Bibliography on Line Shapes and Shifts* covers the literature through June 1978 in four parts [Fuhr *et al.* 1972, 1974, 1975, 1978a]. Bibliographic reports covering line broadening data of particular interest for astronomy as published up through mid-1990 are given in the triennial IAU Transactions [McNally 1991]. A twice-yearly bulletin on atomic and molecular data for fusion [Smith 1991] includes references for line-broadening papers.

The proceedings of a series of conferences published in successive volumes under the title *Spectral Line Shapes* [Frommhold and Keto 1990] are a helpful source of information on research in this area.

## V. Databases and Data Files

Several of the larger and more inclusive collections of computer-readable atomic data are briefly described here, the name of at least one person associated with each database being given. It may reasonably be assumed that all of the larger data centers also maintain bibliographic databases; a few bibliographic databases are mentioned explicitly. It should also be noted that practically all of the larger published collections of atomic spectroscopic data still used extensively by astronomers exist somewhere in computer-readable form. An example from the older literature is the incorporation of Moore's 1945 multiplet tables, with additional material, into a computer-readable line list by Gulliver and Stadel [1990] (see also Adelman *et al.* [1985]). The Standard Reference Data office at NIST issues selected atomic data tables as computer-readable data files or databases (J. Gallagher). Information about computer-readable data tables can be obtained from the pertinent data centers or, in the case of more recent publications, from the authors involved.

The Atomic Energy Levels Data Center and the Data Center on Atomic Transition Probabilities at NIST are building an atomic spectroscopic database for astronomy. This database now includes energy-level, wavelength, and transition-probability data mainly from the more recent compilations published by the two NIST Centers, as described in Parts II and III above (W. C. Martin, W. L. Wiese). One planned means of access will be through the NASA Astrophysics Data System.

New transition-probability data for spectra of the first ten elements now being compiled and entered into the NIST database originate mainly from calculations carried out by participants in the Opacity Project [Seaton 1987, Mendoza 1992]. This project has also included calculations of photoionization cross sections for atoms and ions of these elements. Data from the Opacity Project are maintained in databases in several locations, including the Dept. of Astronomy at Ohio State University, Columbus, OH (A. Pradhan); the Database on Atomic and Molecular Physics at the Queen's University of Belfast (K. A. Berrington, F. J. Smith); and the Scientific Center at IBM de Venezuela, Caracas (C. Mendoza).

R. L. Kurucz of the Harvard-Smithsonian Center for Astrophysics, Cambridge, MA, is revising and extending his computer files of spectroscopic data for stellar atmospheric modeling [Kurucz 1991]. His data on wavelengths, gf values, and damping constants include experimental or experimentally based wavelengths for 555,000 atomic lines. Some 410,000 of these lines belong to the first through the ninth (I through IX) spectra of the iron-group elements calcium through nickel; Kurucz has calculated data for 42 million lines of these elements.

A database on multiply-charged atomic ions at the National Research Institute for Physical-Technical and Radio-Technical Measurements (VNIIFTRI), Mendeleevo, Moscow Region, includes data on energy levels, wavelengths, transition probabilities, electron-impact excitation and ionization, and dielectronic recombination. The coverage of the wavelength region below 100 Å is extensive (A. Ya. Faenov).

A new bibliographic database on atomic spectral line shapes and shifts will be maintained jointly by the Département d'Astrophysique Stellaire et Galactique of the Observatoire de Paris-Meudon (A. Lesage) and by the NIST Data Center on Atomic Line Shapes and Shifts (J. Fuhr). Their plans include publication of annotated bibliographies.

In addition to photoexcitation and photoionization, the Belfast group (see above) has emphasized electron impact excitation data for atoms and ions of astrophysical interest. The Atomic Collisions Database at the Joint Institute for Laboratory Astrophysics, Boulder, CO, also has extensive data of these types (J. Broad).

Heavy particle collisions are emphasized by the Controlled Fusion Atomic Data Center in Oak Ridge, TN (R. A. Phaneuf), and by two Japanese centers: the Data and Planning Center of the National Institute of Fusion Science, Nagoya (H. Tawara), and the Atomic and Molecular Data Unit of the Japan Atomic Energy Research Institute, Tokai-mura (T. Shirai).

The Centre de Donneés GAPHYOR, Université de Paris-Sud, Orsay, operates an extensive bibliographic database covering the types of data mentioned above and also data on molecules, clusters, and interactions of atoms and molecules with solids (J.-L. Delcroix). The assignments of new references into the highly structured system are listed in successive issues of a quarterly bulletin entitled *GAPHYOR Update*.

The Atomic and Molecular Data Center Network of the International Atomic Energy Agency (IAEA), Vienna, comprises fifteen data centers, including most of those mentioned in this article. An increasing collection of data supplied by various Network centers is maintained in a uniform format by the Atomic and Molecular Data Unit of the IAEA (R. K. Janev). The previously mentioned bibliographic bulletin issued twice yearly by this Unit includes references for data on photoionization and electron-impact and heavy-particle collisions [Smith 1991]; retrievals can be made from a bibliographic database extending back to 1950. The IAEA Unit also prepares annual reports that include useful summaries of recent activities of the Network data centers [see, e.g., Janev 1990].

The author would like to thank J. R. Fuhr and A. Musgrove for bibliographic help in the preparation of this article and B. J. DeBord for her careful work on the camera-ready manuscript.

## References

Adam, J., Baschek, B., Johansson, S., Nilsson, A. E. and Brage, T. 1987, Astrophys. J. **312**, 337–343.

Adelman, C. J., Adelman, S. J., Fischel, D. and Warren, Jr., W. H. 1985, Astron. Astrophys., Suppl. Ser. **60**, 339–341

Allard, N. and Kielkopf, J. 1982, Rev. Mod. Phys. **54**, 1103–1182.

Aymar, M., Champeau, R. J., Delsart, C. and Robaux, O. 1984, J. Phys. B **17**, 3645–3661.

Bard, A., Kock, A. and Kock, M. 1991, Astron. Astrophys. **248**, 315–322.

Biémont, E. 1990, see Hansen (Ed.), 48–58.

Biémont, E., Baudoux, M., Kurucz, R. L., Ansbacher, W. and Pinnington, E. H. 1991, Astron. Astrophys. **249**, 539–544.

Biémont, E., Brault, J. W., Delbouille, L. and Roland, G. 1985a, Astron. Astrophys., Suppl. Ser. **61**, 107–125; 1985b, *ibid.* 185–190; 1986, *ibid.* **65**, 21–25.

Biémont, E. and Hansen, J. E. 1986, Phys. Scr. **34**, 116–130.

Bitter, M., Hsuan, H., Decaux, V., Grek, B., Hill, K. W., Hulse, R., Kruegel, L. A., Johnson, D., von Goller, S. and Zarnstorff, M. 1991, Phys. Rev. A **44**, 1796–1805.

Blackwell, D. E. 1990, see Hansen (Ed.), 160–164.

Blaise, J., Vergès, J., Wyart, J.-F. and Engleman, Jr., R. 1992, J. Phys. II (Paris) **2**, to be published.

Blaise, J. and Wyart, J.-F. 1992a, *Energy Levels and Atomic Spectra of Actinides*, International Tables of Selected Constants **20**, Tables de Constantes, Paris.

Blaise, J. and Wyart, J.-F. 1992b, J. Res. Natl. Inst. Stand. Tech. **97**, 217–223.

Blaise, J., Wyart, J.-F., Djerad, M. T. and Ahmed, Z. B. 1984, Phys. Scr. **29**, 119–131.

Boiko, V. A., Pal'chikov, V. G., Skobelev, I. Yu. and Faenov, A. Ya. 1988, *X-Ray Spectroscopy of Multicharged Ions* (in Russian), Energoatomizdat, Moscow.

Bridges, J. M. and Kornblith, R. L. 1974, Astrophys. J. **192**, 793–812.

Brooks, N. H., Rohrlich, D. and Smith, W. H. 1977, Astrophys. J **214**, 328–330.

Brown, C. M., Ginter, M. L., Johansson, S. and Tilford, S. G. 1988, J. Opt. Soc. Am. B **5**, 2125–2158.

Buchet, J. P., Buchet-Poulizac, M. C., Denis, A., Desesquelles, J., Druetta, M., Martin, S. and Wyart, J.-F. 1987, J. Phys. B **20**, 1709–1723.

Chang, M.-W. 1977, Astrophys. J. **211**, 300–307.

Chen, M. H. 1986, At. Data Nucl. Data Tables **34**, 301–356.

Cooksy, A. L., Blake, G. A. and Saykally, R. J. 1986a, Astrophys. J. **305**, L89–L92

Cooksy, A. L., Hovde, D. C. and Saykally, R. J. 1986b, J. Chem. Phys. **84**, 6101–6107.

Cooksy, A. L., Saykally, R. J., Brown, J. M. and Evenson, K. M. 1986c, Astrophys. J. **309**, 828–832.

Copeland, R. A., Jeffries, J. B., Hickman, A. P. and Crossley, D. R. 1987, J. Chem. Phys. **86**, 4876–4884.

Corliss, C. H. and Bozman, W. R. 1962, *Experimental Transition Probabilities for Spectral Lines of Seventy Elements*, Natl. Bur. Stand. (U.S.) Mono. **53**.

Cowley, C. R. and Frey, M. 1988, Nucl. Instrum. Meth. Phys. Res. B **31**, 214–221.

Crosswhite, H. M. 1975, J. Res. Natl. Bur. Stand. **79A**, 17–69.

Dimitrijević, M. S. 1988, Astron. Astrophys., Suppl. Ser. **76**, 53–59.

Dimitrijević, M. S. 1990, in Wehrse, R. (Ed.), *Accuracy of Element Abundances from Stellar Atmospheres*, Lecture Notes in Physics **356**, Springer-Verlag, Berlin, 31–44.

Dimitrijević, M. S., Sahal-Bréchot, S. and Bommier, V. 1991a, Astron. Astrophys., Suppl. Ser. **89**, 581–590; 1991b, *ibid.*, **89**, 591–598.

Drake, G. W. F. 1988, Can. J. Phys. **66**, 586–611.

Edlén, B. 1978, Phys. Scr. **17**, 565–574; 1979, *ibid.* **19**, 255–266; 1983a, *ibid.* **28**, 51–67; 1983b, *ibid.* **28**, 483–495; 1984, *ibid.* **30**, 135–145; 1985a, *ibid.* **31**, 345–358; 1985b, *ibid.* **32**, 86–88.

Ekberg, J. O., Smitt, R., Skogvall, B. and Borgström, A. 1990, Phys. Scr. **41**, 217–226.

Engleman, Jr., R. 1985, J. Opt. Soc. Am. B **2**, 1934–1941.

Engleman, Jr., R. 1989, Astrophys. J. **340**, 1140–1143.

Erickson, G. W. 1977, J. Phys. Chem. Ref. Data **6**, 831–869.

Eriksson, K. B. S. 1986, Phys. Scr. **34**, 211–215.

Eriksson, K. B. S. 1987, J. Opt. Soc. Am. B **4**, 1369–1371.

Fawcett, B. C. 1984, Phys. Scr. **30**, 326–334.

Forsberg, P. 1991, Phys. Scr. **44**, 446–476.

Forsberg, P., Johansson, S. and Smith, P. L. 1986, Phys. Scr. **34**, 759–765.

Frommhold, L. and Keto, J. W. (Eds.) 1990, *Spectral Line Shapes*, Vol. 6, AIP Conf. Proc. 216, Amer. Inst. Phys., New York; and references for Vols. 1–5 therein.

Fuhr, J. R., Martin, G. A. and Specht, B. J. 1975, *Bibliography on Atomic Line Shapes and Shifts (July 1973 through May 1975)*, Natl. Bur. Stand. (U.S.), Spec. Publ. 366, Suppl. 3.

Fuhr, J. R., Martin, G. A. and Wiese, W. L. 1988, *Atomic Transition Probabilities–Iron through Nickel*, J. Phys. Chem. Ref. Data **17**, Suppl. 4.

Fuhr, J. R., Miller, B. J. and Martin, G. A. 1978a, *Bibliography on Atomic Line Shapes and Shifts (June 1975 through June 1978)*, Natl. Bur. Stand. (U.S.), Spec. Publ. 366, Suppl. 3.

Fuhr, J. R., Miller, B. J. and Martin, G. A. 1978b, *Bibliography on Atomic Transition Probabilities (1914 through October 1977)*, Natl. Bur. Stand. (U.S.), Spec. Publ. 505.

Fuhr, J. R., Roszman, L. J. and Wiese, W. L. 1974, *Bibliography on Atomic Line Shapes and Shifts (April 1972 through June 1973)*, Natl. Bur. Stand. (U.S.), Spec. Publ. 366, Suppl. 1.

Fuhr, J. R. and Wiese, W. L. 1990, *Atomic Transition Probabilities*, in Lide, D. R. (Ed.), *CRC Handbook of Chemistry and Physics*, 71st Edition, CRC Press, Boca Raton, FL.

Fuhr, J. R., Wiese, W. L. and Roszman, L. J. 1972, *Bibliography on Atomic Line Shapes and Shifts (1889 through March 1972)*, Natl. Bur. Stand. (U.S.), Spec. Publ. 366.

Garcia-Riquelme, O. and Rico, F. R. 1991, *Espectro Ni III*, Instituto de Optica "Daza de Valdes" (CSIC), Madrid, Publ. No. 49.

Ginibre, A. 1981, Phys. Scr. **23**, 260–267; 1989a, *ibid.* **39**, 694–709; 1989b, *ibid.* 710–721.

Ginibre, A. 1990, At. Data Nucl. Data Tables **44**, 1–29.

Goldbach, C., Martin, M., Nollez, G., Plomdeur, P., Zimmermann, J. P. and Babic, D. 1986, Astron. Astrophys. **161**, 47–54.

Goldbach, C. and Nollez, G. 1988, Astron. Astrophys. **201**, 189–193.

Gulliver, A. F. and Stadel, J. G 1990, Publ. Astron. Soc. Pac. **102**, 587–591.

Hagan, L. 1977, *Bibliography on Atomic Energy Levels and Spectra, July 1971 through June 1975*, Natl. Bur. Stand. (U.S.), Spec. Publ. 363, Suppl. 1.

Hagan, L. and Martin, W. C. 1972, *Bibliography on Atomic Energy Levels and Spectra, July 1968 through June 1971*, Natl. Bur. Stand. (U.S.), Spec. Publ. 363.

Hansen, J. E. (Ed.) 1990, *Atomic Spectra and Oscillator Strengths for Astrophysics and Fusion Research*, North-Holland, Amsterdam.

Hansen, J. E. and Persson, W. 1987, Phys. Scr. **36**, 602–643.

Hata, J. and Grant, I. P. 1984, Mon. Not. R. Astron. Soc. **211**, 549–557.

Heilig, K. 1977, Spectrochim. Acta **32B**, 1–57; 1982, *ibid.* **37B**, 417–455; 1987, *ibid.* **42B**, 1237–1266; 1992, *ibid.* **47B**, 303–326.

Hibbert, A. 1990, see Hansen (Ed.), 102–107.

Hibbert, A., Biémont, E., Godefroid, M. and Vaeck, N. 1991, Astron. Astrophys., Suppl. Ser. **88**, 505–524.

Iglesias, L., Cabeza, M. I., and de Luis, B. 1988, *The Spectrum of the $V^+$ Ion (V II)*, Instituto de Optica "Daza de Valdes" (CSIC), Madrid, Publ. No. 47.

Iglesias, L., Cabeza, M. I., Rico, F. R., Garcia-Riquelme, O. and Kaufman, V. 1989, J. Res. Natl. Inst. Stand. Tech. **94**, 221–258

Isberg, B. and Litzén, U. 1985, Phys. Scr. **31**, 533–538; 1986, *ibid.* **33**, 420–423.

Janev, R. K. (Ed.) 1990, *Summary Report, IAEA Consultant's Meeting: 9th Meeting of the Atomic and Molecular Data Centres and ALADDIN Network*, INDC (NDS)–243/M7, International Atomic Energy Agency, Vienna.

Johansson, S. 1991, *High-Resolution Atomic Spectroscopy in the Infrared and Its Application to Astrophysics*, in Jaschek, C. and Andrillat, Y. (Eds.), *The Infrared Spectral Region of Stars*, Cambridge Univ. Press, Cambridge.

Johansson, S. and Baschek, B. 1988, Nucl. Instrum. Meth. Phys. Res. B **31**, 222–232.

Johansson, S. and Cowley, C. R. 1988, J. Opt. Soc. Am. B **5**, 2264–2279.

Johansson, S. and Learner, R. C. M. 1990, Astrophys. J. **354**, 755–762.

Johansson, S., Nave, G., Learner, R. C. M. and Thorne, A. P. 1991a, *New Identifications of Highly-Excited Fe I Lines around 4μm in Cool Stars*, in *The Infrared Spectral Region of Stars*, see Johansson 1991.

Johansson, S., Rosberg, M. and Hamann, F. 1991b, *Identification of New 14 eV Fe II Lines in the Near-Infrared Spectrum of LkHα 101*, in *The Infrared Spectral Region of Stars*, see Johansson 1991.

Johnson, W. R. and Soff, G. 1985, At. Data Nucl. Data Tables **33**, 405–446.
Joshi, Y. N., Raassen, A. J. J. and Arcimowicz, B. 1989, J. Opt. Soc. Am. B **6**, 534–538.
Karim, K. R. and Bhalla, C. P. 1990, Phys. Rev. A **42**, 3555–3560.
Jupén, C., Litzén, U., Kaufman, V. and Sugar, J. 1987, Phys. Rev. A **35**, 116–130.
Karim, K. R. and Bhalla, C. P. 1990, Phys. Rev. A **42**, 3555–3560.
Kaufman, V. and Edlén, B. 1974, J. Phys. Chem. Ref. Data **3**, 825–895.
Kaufman, V. and Martin, W. C. 1991a, J. Phys. Chem. Ref. Data **20**, 83–152; 1991b, *ibid.* 775–858.
Kaufman, V. and Sugar, J. 1986, J. Phys. Chem. Ref. Data **15**, 321–426; 1988, *ibid.* **17**, 1679–1789.
Kelly, R. L. 1979, *Atomic Emission Lines in the Near Ultraviolet; Hydrogen through Krypton*, TM 80268, National Aeronautics and Space Administration, Goddard Space Flight Center, Greenbelt, MD.
Kelly, R. L. 1987, *Atomic and Ionic Spectrum Lines below 2000 Angstroms: Hydrogen through Krypton*, J. Phys. Chem. Ref. Data **16**, Suppl. 1.
Kim, Y.-K., Baik, D. H., Indelicato, P. and Desclaux, J. P. 1991, Phys. Rev. A **44**, 148–166.
Kim, Y.-K., Martin, W. C. and Weiss, A. W. 1988, J. Opt. Soc. Am. B **5**, 2215–2224.
Konjević, N., Dimitrijević, M. S. and Wiese, W. L. 1984a, J. Phys. Chem. Ref. Data **13**, 619–647; 1984b, *ibid.* 649–686.
Konjević, N. and Roberts, J. R. 1976, J. Phys. Chem. Ref. Data **5**, 209–257.
Konjević, N. and Wiese, W. L. 1976, J. Phys. Chem. Ref. Data **5**, 259–308; 1990, *ibid.* **19**, 1307–1385.
Kurucz, R. L. 1981, *Semiempirical Calculation of gf Values, IV: Fe II*, Smithsonian Astrophysical Observatory Report **390**, Cambridge, MA.
Kurucz, R. L. 1991, "New Opacity Calculations", in Crivellari, L., Hubeny, I. and Hummer, D. G. (Eds.), *Stellar Atmospheres: Beyond Classical Models*, NATO ASI Series, Kluwer Academic, Dordrecht, 440–448.
Kurucz, R. L. and Peytremann, E. 1975, *A Table of Semiempirical gf Values*, Parts 1–3, Smithsonian Astrophysical Observatory Special Report **362**, Cambridge, MA.
Learner, R. C. M. and Thorne, A. P. 1988, J. Opt. Soc. Am. B **5**, 2045–2059.
Leckrone, D. S., Johansson, S., Kurucz, R. L. and Adelman, S. J. 1990, see Hansen (Ed.), 3–12.
Litzén, U. 1990, see Hansen (Ed.), 198–199.

Martin, G. A., Fuhr, J. R. and Wiese, W. L. 1988, *Atomic Transition Probabilities–Scandium through Manganese*, J. Phys. Chem. Ref. Data **16**, Suppl. 3.

Martin, W. C. 1981, Phys. Scr. **24**, 725–731.

Martin, W. C. 1987, Phys. Rev. A **36**, 3575–3589.

Martin, W. C. and Zalubas, R. 1979, J. Phys. Chem. Ref. Data **8**, 817–864; 1980, *ibid.* **9**, 1–58; 1981, *ibid.* **10**, 153–195; 1983, *ibid.* **12**, 323–380.

Martin, W. C., Zalubas, R. and Hagan, L. 1978, *Atomic Energy Levels–The Rare-Earth Elements*, Natl. Stand. Ref. Data Ser., Natl. Bur. Stand. (U.S.) **60**.

Martin, W. C., Zalubas, R. and Musgrove, A. 1985, J. Phys. Chem. Ref. Data **14**, 751–802; 1990, *ibid.* **19**, 821–880.

Martinson, I. 1989, Rep. Prog. Phys. **52**, 157–225.

Martinson, I. 1990, see Hansen (Ed.), 153–159.

Mathews, C. W., Ginter, M. L., Ginter, D. S. and Brown, C. M. 1989, J. Opt. Soc. Am. B **6**, 1627–1643.

Mathys, G. 1985, Astron. Astrophys., Suppl. Ser. **59**, 229–253.

Mathys, G. 1988, in Adleman, S. J. and Lanz, T. (Eds.), *Elemental Abundance Analyses*, Inst. d'Astron. de l'Université de Lausanne, 143–147.

Mazzoni, M., Joshi, Y. N., Nencioni, A., Grisendi, T. and Parkinson, W. H. 1987, J. Phys. B **20**, 2193–2202.

McNally, D. (Ed.) 1991, *Reports on Astronomy: Trans. Int. Astron. Union XXIA*, Kluwer Academic, Dordrecht, see *Commission 14: Atomic and Molecular Data*, 105–136; and reports of Commission 14 in preceding volumes of Trans. Int. Astron. Union.

Meggers, W. F., Corliss, C. H. and Scribner, B. F. 1975, *Tables of Spectral-Line Intensities* (Part I–Arranged by Elements, Part II–Arranged by Wavelengths), 2nd Edition, Natl. Bur. Stand. (U.S.), Monogr. 145.

Meggers, W. F. and Tech, J. L. 1978, J. Res. NBS **83**, 13–70.

Mendoza, C. 1992, chapter entitled "Atomic Data from the Opacity Project" in this volume.

Miller, B. J., Fuhr, J. R. and Martin, G. A. 1980, *Bibliography on Atomic Transition Probabilities (November 1977 through March 1980)*, Natl. Bur. Stand. (U.S.), Spec. Publ. 505, Suppl 1.

Mohr, P. J. 1983, At. Data Nucl. Data Tables **29**, 453–466.

Moore, C. E. 1945, *A Multiplet Table of Astrophysical Interest*, Contrib. Princeton Univ. Obs. No. 20; reprinted 1972, Natl. Stand. Ref. Data Ser., Natl. Bur. Stand. (U.S.) **40**.

Moore, C. E. 1949, *Atomic Energy Levels*, Natl. Bur. Stand. (U.S.), Circ. 467, Vol. I; 1952, Vol. II; 1958, Vol. III; reprinted 1971, Natl. Stand. Ref. Data Ser., Natl. Bur. Stand. (U.S.) **35**.

Moore, C. E. 1950, *An Ultraviolet Multiplet Table*, Natl. Bur. Stand. (U.S.), Circ. 488, Sec. 1; 1952, Sec. 2; 1962, Secs. 3, 4, 5; reprinted 1968.

Moore, C. E. 1965, *Selected Tables of Atomic Spectra*, Natl. Stand. Ref. Data Ser., Natl. Bur. Stand. (U.S.) **3**, Sect. 1 (Si II–IV); 1967, Sec. 2 (Si I); 1970b, Sec. 3 (C I–VI); 1971, Sec. 4 (N IV–VII); 1975, Sec. 5 (N I–III); 1972, Sec. 6 (H I, D, T); 1976, Sec. 7 (O I); 1979, Sec. 8 (O VI–VIII); 1980, Sec. 9 (O V); 1982, Sec. 10 (O IV); 1985, Sec. 11 (O III).

Moore, C. E. 1968, *Bibliography on the Analyses of Optical Atomic Spectra*, Natl. Bur. Stand. (U.S.), Spec. Publ. 306, Sec. 1; 1969a, Sec. 2; 1969b, Sec. 3; 1969c, Sec. 4.

Moore, C. E. 1970a, *Ionization Potentials and Ionization Limits Derived from the Analyses of Optical Spectra*, Natl. Stand. Ref. Data Ser., Natl. Bur. Stand. (U.S.) **34**.

Morton, D. C. 1991, Astrophys. J., Suppl. Ser. **77**, 119–202.

Mount, G. H., Yamasaki, G., Fowler, W. and Fastie, W. G. 1977, Appl. Opt. **16**, 591–595.

Musgrove, A. and Zalubas, R. 1985, *Bibliography on Atomic Energy Levels and Spectra, July 1979 through December 1983*, Natl. Bur. Stand. (U.S.), Spec. Publ. 363, Suppl. 3.

Nave, G., Learner, R. C. M., Thorne, A. P. and Harris, C. J. 1991, J. Opt. Soc. Am. B **8**, 2028–2041.

Nilsen, J. 1988, At. Data Nucl. Data Tables **38**, 339–379.

Nilsson, A. E., Johansson, S. and Kurucz, R. L. 1991, Phys. Scr. **44**, 226–257.

O'Brian, T. R. and Lawler, J. E. 1991, Phys. Rev. A **44**, 7134–7143.

O'Brian, T. R., Wickliffe, M. E., Lawler, J. E., Whaling, W. and Brault, J. W. 1991, J. Opt. Soc. Am. B **8**, 1185–1201.

Odintzova, G. A. and Striganov, A. R. 1979, J. Phys. Chem. Ref. Data **8**, 63–67.

Outred, M. 1978, J. Phys. Chem. Ref. Data **7**, 1–262.

Palmer, B. A. and Engleman, Jr., R. 1983, Los Alamos Natl. Lab. Report LA-9615, UC-4, Los Alamos Natl. Lab., Los Alamos, NM 87545.

Palmer, B. A., Keller, R. A. and Engleman, Jr., R. 1980, Los Alamos Natl. Lab. Report LA-8251-MS, UC-34a, Los Alamos Natl. Lab., Los Alamos NM 87545.

Persson, W., Wahlström, C. G., Jönsson, L. and Di Rocco, H. O. 1991, Phys. Rev. A **43**, 4791–4823.

Pettersson, S.-G. and Wenåker, I. 1990, Phys. Scr. **42**, 187–191.

Raassen, A. J. J., van der Valk, A. A. and Joshi, Y. N. 1990, see Hansen (Ed.), 202–205.

Reader, J., Acquista, N., Sansonetti, C. J. and Engleman, Jr., R. 1988, J. Opt. Soc. Am. B **5**, 2106–2118.

Reader, J., Acquista, N., Sansonetti, C. J. and Sansonetti, J. E. 1990a, Astrophys. J., Suppl. Ser. **72**, 831–866.

Reader, J. and Corliss, C. H. 1990, *Line Spectra of the Elements*, in Lide, D. R. (Ed.), *CRC Handbook of Chemistry and Physics*, 71st Edition, CRC Press, Boca Raton, FL.

Reader, J., Corliss, C. H., Wiese, W. L. and Martin, G. A. 1980, *Wavelengths and Transition Probabilities for Atoms and Atomic Ions* (Part I–Wavelengths, Part II–Transition Probabilities), Natl. Stand. Ref. Data Ser., Natl. Bur. Stand. (U.S.) **68**.

Reader, J., Ekberg, J. O., Feldman, U., Brown, C. M. and Seely, J. F. 1990b, J. Opt. Soc. Am. B **7**, 1176–1181.

Reader, J., Kaufman, V., Sugar, J., Ekberg, J. O., Feldman, U., Brown, C. M., Seely, J. F. and Rowan, W. L. 1987, J. Opt. Soc. Am. B **4**, 1821–1828.

Rosberg, M. and Johansson, S. 1992, Phys. Scr. (submitted 1991).

Ryabtsev, A. N. and Wyart, J.-F. 1987, Phys. Scr. 255–261.

Safronova, U. I. and Lisina, T. G. 1979, At. Data Nucl. Data Tables **24**, 49–93.

Sansonetti, C. J. and Weber, K. H. 1984, J. Opt. Soc. Am. B **1**, 361–365.

Sansonetti, J. E., Reader, J., Sansonetti, C. J. and Acquista, N. 1992, J. Res. Natl. Inst. Stand. Tech. **97**, 1–212.

Seaton, M. J. 1987, J. Phys. B **20**, 6363–6378.

Seely, J. F., Brown, C. M., Feldman, U., Ekberg, J. O., Keane, C. J., MacGowan, B. J., Kania, D. R. and Behring, W. E. 1991, At. Data Nucl. Data Tables **47**, 1–15.

Seely, J. F., Feldman, U. and Safronova, U. I. 1986, Astrophys. J. **304**, 838–848.

Shirai, T., Funatake, Y., Mori, K., Sugar, J., Wiese, W. L., and Nakai, Y. 1990, J. Phys. Chem. Ref. Data **19**, 127–275.

Shirai, T., Mengoni, A., Nakai, Y., Sugar, J., Wiese, W. L., Mori, K., and Sakai, H. 1992a, J. Phys. Chem. Ref. Data **21**, 23–121.

Shirai, T., Mori, K., Sugar, J., Wiese, W. L., Nakai, Y. and Ozawa, K. 1987a, At. Data Nucl. Data Tables **37**, 235–332.

Shirai, T., Nakagaki, T., Nakai, Y., Sugar, J., Ishii, K., and Mori, K. 1991, J. Phys. Chem. Ref. Data **20**, 1–81.

Shirai, T., Nakagaki, T., Sugar, J. and Wiese, W. L. 1992b, J. Phys. Chem. Ref. Data **21** (in press).

Shirai, T., Nakai, Y., Nakagaki, T., Sugar, J. and Wiese, W. L. 1992c, J. Phys. Chem. Ref. Data **21** (to be submitted).

Shirai, T., Nakai, Y., Ozawa, K., Ishii, K., Sugar, J. and Mori, K. 1987b, J. Phys. Chem. Ref. Data **16**, 327–377.

Smith, J. J. (Ed.) 1991, *International Bulletin on Atomic and Molecular Data for Fusion*, No. 42, International Atomic Energy Agency, Vienna; and preceding issues.

Smith, P. L., Huber, M. C. E., Tozzi, G. P., Giresinger, H. E., Cardon, B. L. and Lombardi, G. G. 1987, Astrophys. J. **322**, 573–583.

Striganov, A. R. and Odintsova, G. A. 1982, *Tables of Spectral Lines of Atoms and Ions* (in Russian), Energoizdat, Moscow.

Striganov, A. R. and Sventitskii, N. S. 1968, *Tables of Spectral Lines of Neutral and Ionized Atoms*, IFI/Plenum, New York.

Sugar, J. and Corliss, C. 1985, *Atomic Energy Levels of the Iron-Period Elements: Potassium through Nickel*, J. Phys. Chem. Ref. Data **14**, Suppl. 2.

Sugar, J. and Kaufman, V. 1979, J. Opt. Soc. Am. **69**, 141–143.

Sugar, J., Kaufman, V. and Rowan, W. L. 1992, J. Opt. Soc. Am. B **9**, 344–346.

Sugar, J. and Musgrove, A. 1988, J. Phys. Chem. Ref. Data **17**, 155–239; 1990, *ibid.* **19**, 527–616; 1991, *ibid.* **20**, 859–915.

Suskin, M. and Weiss, A. W. 1992, to be submitted.

Thévenin, F. 1989, Astron. Astrophys., Suppl. Ser. **77**, 137–154; 1990, *ibid.* **82**, 179–188.

Vainshtein, L. A. and Safronova, U. I. 1978, At. Data Nucl. Data Tables **21**, 49–68; 1980, *ibid.* **25**, 311–385; 1985, Phys. Scr. **31**, 519–532.

van der Valk, A. A., Raassen, A. J. J. and Joshi, Y. N. 1990, J. Opt. Soc. Am. B **7**, 1182–1189.

Wenåker, I. 1990, Phys. Scr. **42**, 667–684.

Wiese, W. L. and Fuhr, J. R. 1990, in Wehrse, R. (Ed.), *Accuracy of Element Abundances from Stellar Atmospheres*, Lecture Notes in Physics **356**, Springer-Verlag, Berlin, 7–18.

Wiese, W. L., Smith, M. W and Glennon, B. M. 1966, *Atomic Transition Probabilities, Volume I, Hydrogen through Neon*, Natl. Stand. Ref. Data Ser., Natl. Bur. Stand. (U.S.) **4**.

Wiese, W. L., Smith, M. W. and Miles, B. M. 1969, *Atomic Transition Probabilities, Volume II, Sodium through Calcium*, Natl. Stand. Ref. Data Ser., Natl. Bur. Stand. (U.S.) **22**.

Wilson, M. 1990, Phys. Lett. A **147**, 215–217.

Wyart, J.-F. 1985, Phys. Scr. **32**, 58–63.

Wyart, J.-F. and Camus, P. 1979, Phys. Scr. **20**, 43–59.

Wyart, J.-F., Raassen, A. J. J., Joshi, Y. N. and Uylings, P. H. M. 1992, J. Phys. II (Paris) **2**, to be published.

Zaidel', A. N., Prokof'ev, V. K., Raiskii, S. M., Slavnyi, V. A. and Shreider, E. Ya. 1970, *Tables of Spectral Lines*, IFI/Plenum, New York. An updated (fourth) edition was published in Russian; 1977, *ibid.*, Nauka, Moscow.

Zalubas, R. and Albright, A. 1980, *Bibliography on Atomic Energy Levels and Spectra, July 1975 through June 1979*, Natl. Bur. Stand. (U.S.), Spec. Publ. 363, Suppl. 2.

Zeippen, C. J. 1990, see Hansen (Ed.), 114–119.

Zhu, Q., Bridges, J. M., Hahn, T. and Wiese, W. L. 1989, Phys. Rev. A **40**, 3721–3726.

Zink, L. R., Evenson, K. M., Matsushima, F., Nelis, T. and Robinson, R. L. 1991, Astrophys. J. **371**, L85–L86.

# 9

# Summary of Current Molecular Databases

W. H. Parkinson[1]

ABSTRACT
There are a large number of atlases, bibliographies, books, and monographs, along with their vital revisions and new editions, that form the foundation of modern, computer-readable, molecular databases. A brief summary is given of some of the important books, current compilations and databases. Included, for example, are the great compilations, *The Identification of Molecular Spectra* by Pearse and Gaydon, and *Molecular Spectra and Molecular Structure* by Huber and Herzberg, and other definitive reference material, circulars, and monographs from many laboratories. These works have laid the foundation for, and have become part of, current databases such as HITRAN, the JPL catalogue, and RADEN, which support studies of, and spectral synthesis programs for, the transmission and emission of radiation in planetary and stellar atmospheres.

## Introduction

Compilations of molecular parameters have grown with the needs and support of physicists and astronomers. Specific requirements for spectroscopic data have come from research projects in fusion, planetary atmospheres, and astronomy. Often these compilations of spectroscopic data are used to support spectral synthesis programs. The content, format, and aim of compilations (or databases) are varied and reflect the current knowledge and needs of research. Because of the nature of this subject, and space limitations, only a brief summary of important compilations has been included here.

There are a large number of atlases, bibliographies, books, and monographs that form the foundation of modern molecular databases. Among such original reference material, one would count: *Données Spectroscopique* edited by B. Rosen (1951, 1970), *The Berkeley Analysis of Molecular Spectra* by J. G. Phillips and S. Davis (1963, 1968, 1987), the bi-monthly *Berke-*

[1] Harvard-Smithsonian Center for Astrophysics, 60 Garden Street, Cambridge, MA 02138, USA

*ley Newsletter Analysis of Molecular Spectra*, the set of *Identification Atlases of Molecular Spectra* (Series 1, 1964-1972 and Series 2, 1992 ) by R. W. Nicholls, and various circulars and monographs published by the National Bureau of Standards and the U.S. Government Printing Office. For example in the 1960's, spectroscopic work in the infrared on atmospheric molecular features, such as the water vapor bands in the 2.7 $\mu$m region (Gates *et al.* 1964), the carbon dioxide bands in the 2.0-2.7 $\mu$m region (Colfee and Benedict 1966), and the $CH_4$ bands near 3 and 7.5 $\mu$m (Kyle 1968), became part of the base for the Air Force Cambridge Research Laboratory atmospheric absorption line parameter compilation (McClatchey *et al.* 1973). This computer-readable database was a compilation of spectroscopic data for the seven significant, infrared-active gases in the Earth's atmosphere. The 'AFGL-tape', as it was known, has been periodically updated by L. S. Rothman and colleagues since its first release in 1973 (Rothman and McClatchey 1976). It is the basis for a number of other compilations, and is now known as the high-resolution transmission molecular absorption (HITRAN) database (see later section).

In the next sections, a summary of selected books, compilations, and databases is given, and where possible, the name of the manager of each.

## Current Molecular Compilations and Databases

*The Identification of Molecular Spectra* by R.W.B. Pearse and A. G. Gaydon (1976, reprinted 1984) is still extant. This compilation, first published in 1941, continues to be unique and indispensable for identification of molecular spectra (both diatomic and polyatomic) and for information on the appearance and occurrence of each band spectrum. There are data on the spectra of more than 490 diatomic and 127 polyatomic molecules and twelve plates of about 60 spectra.

*Molecular Spectra and Molecular Structure, IV, Constants of Diatomic Molecules* by K. P Huber and G. Herzberg (1979) is a critical and comprehensive compilation of constants of more than 900 diatomic molecules. It is a 700 page table of electronic states, symmetry notation, electronic energy, vibrational and rotational constants, internuclear distance, observed transitions, their $\nu_{00}$ values, references, and many informative footnotes. To date there is only one edition of this unique book. One hopes that the means will be found to continue this compilation with revisions and new editions.

A compilation titled *Absolute Cross Sections for Molecular Photoabsorption, Partial Photoionization, and Ionic Photofragmentation Processes* has been critically assembled and evaluated by Gallagher *et al.* (1988). These data are for diatomics ($H_2$, $N_2$, $O_2$, CO, NO), triatomics ($CO_2$, $N_2O$), hydrides ($H_2O$, $NH_3$, $CH_4$), hydrogen halides (HF, HCl, HBr, HI), sulfur compounds ($H_2S$, $CS_2$, OCS, $SO_2$, $SF_6$), and chlorine compounds ($Cl_2$,

$CCl_6$). The data compiled and presented in tables and figures are both experimental and theoretical, and cover a 10 year period to 1986.

A compilation of measurements of photoabsorption and photoionization cross sections of O, $N_2$, and $O_2$ has been assembled by R. R. Conway for use in studies of ion and photoelectron production in the ionosphere (Conway 1988). The main differences from an earlier compilation (Kirby *et al.* 1979) for aeronomical calculations are a factor of 2.3 reduction of the O absorption cross section at 150 Å, and more detailed information on, and understanding of, the dissociating states of $N_2$ and $O_2$. The wavelength range has been matched to the Atmospheric Explorer solar reference wavelength scale of Hinteregger *et al.* (1981) and extends from 18.62 Å to 1050.01 Å. Total and partial cross sections for each species are presented separately in tables; branching ratios for the production of electronic states of the ion are normalized separately for bound and fragmenting states and are presented graphically; partial cross sections are in tables; figures are used to compare laboratory data with the interpolated data in the tables.

A compilation by F. J. Lovas and R. D. Suenram of all of the rotational microwave lines observed and reported in the open literature for 91 hydrocarbons, CH to $C_{10}H_{10}$, have been tabulated (1989). Their isotopic molecular species, assigned quantum numbers, observed frequencies, estimated measurement uncertainties and references are given for each transition reported. The spectral lines for many normal isotopic species have been refit to produce a comprehensive and consistent analysis of all the data extracted from various literature sources, and a number of misprints and errors in the literature have been corrected. The derived molecular properties, such as rotational and centrifugal distortion constants, hyperfine structure constants, electric dipole moments, and rotational *g*-factors, are listed.

In order to provide some of the basic molecular spectroscopic references and data needed to support the Infrared Space Observatory (ISO), J. Crovisier of the Observatoire de Paris - Meudon has started a compilation of microwave and infrared data and references for molecules in gas phase. The compilation is in the initial draft stage but it is intended to include data on most molecules detected in the interstellar medium, in comets, and in planetary atmospheres, but not those in the Earth's atmosphere. A few isotopic species will also be included. As currently planned, the data will not be selected nor will the quality of the data be assessed. Data for about 185 molecules will be collected and summarized for individual species. Each species will be identified by name(s) and formula, electronic ground state, and symmetry. Photodissociation rates, if known, will be given at one AU from the Sun and in the interstellar medium (by assuming the UV radiation field of Mathis *et al.* 1983). Fundamental vibrational numbering, frequency ($cm^{-1}$), band or intensity strengths ($cm^{-2}$ $atm^{-1}$ at STP), Einstein A-coefficient or transition probability ($s^{-1}$), and excitation rate, g ($s^{-1}$), will be listed.

# HITRAN '92 (L. S. Rothman)

In order to support studies of, and spectral synthesis programs for, the transmission and emission of radiation through the Earth's atmosphere, the US Air Force Phillips Laboratory/Geophysics Directorate developed HITRAN. HITRAN '92, the latest edition of the spectroscopic molecular database (the 'AFGL-tape') was released in March 1992 by L. S. Rothman and colleagues and supersedes HITRAN '91 (Rothman *et al.* 1987; Park *et al.* 1987). This most comprehensive compilation contains data on the seven principal atmospheric molecules, twenty-four additional trace molecular species, and major isotopic variants. These molecules have bands covering the wavelength region from 0 to almost 23,000 $cm^{-1}$. There are 647,672 lines in the high-resolution section, as well as cross-section data for many heavy molecular species that cannot be represented on a line-by-line basis, such as chlorofluorocarbons, $SF_6$, and oxides of nitrogen.

For each molecular transition, the database contains the following parameters: molecule number, isotope number, frequency ($cm^{-1}$), line intensity ($cm^{-1}$ molec.$^{-1}$ $cm^{-2}$ at 296 K), transition probability (Debyes)$^2$, air-broadened halfwidth ($cm^{-1}$ $atm^{-1}$ at 296 K), lower state energy ($cm^{-1}$), exponent for the temperature dependence of the air-broadened halfwidth, shift of transition due to pressure, upper state and lower state global quanta indices, upper state and lower state local quanta, accuracy indices for frequency, intensity, and halfwidth, and indices for looking up of references for frequency, intensity, and halfwidth.

In the first issue of the *HITRAN Newsletter* (1991), Rothman reported on the '91 HITRAN, on a high-temperature version (HITEMP), and on some major advances in the creations of PC-oriented interfaces for HITRAN. The PC version of 1991 HITRAN will be sold by the University of South Florida Research Foundation. The software has been changed from sequential access to random access so as to increase the data search speed by two to three orders of magnitude. The data files come on 14 standard floppy diskettes with value-added software.

In the second issue of the *HITRAN Newsletter* (1992), Rothman briefly describes the compilation and alternative media that contain the database. He reports that a special issue of the *Journal of Quantitative Spectroscopy and Radiative Transfer* will present, later in 1992, a detailed summary of HITRAN '92 and sources of data.

# GEISA (N. Husson)

The Laboratoire de Meteorologie Dynamique de CNRS in France released on May 1991 the latest version of the GEISA (Gestion et Étude des Information Spectroscopic Atmosphériques) molecular spectroscopic line database by Husson *et al.* (1991). The GEISA database has always followed very closely the format and content of HITRAN, but with the addition of molecules of particular importance to planetary atmospheres and especially to the giant planets. The most recently announced edition is standardized with HITRAN. There are approximately 720,000 lines in the wavelength region 0 to 22,656 $cm^{-1}$ corresponding to 40 molecules and 86 isotopic species.

## JPL Catalogue (H. M. Pickett, R. L. Poyner and E. A. Cohen)

The computer-accessible spectral line catalogue from the Jet Propulsion Laboratory (JPL) was originally published by H. M. Pickett and R. L. Poynter in 1984 and 1985. The new edition (1992) covers the submillimeter to microwave wavelength range (0-10,000 GHz) and includes more than 630,924 spectral lines. The catalogue contains 206 atomic and molecular species of interest to radio astronomers and atmospheric scientists. New measurements of the $O_3$ spectrum in the microwave and infrared wavelength range (Pickett *et al.* 1985 and Pickett *et al.* 1988) added data on $O_3$ rotational lines, and, in particular, to positions of $\nu_2 + (2\nu_2 - \nu_2)$. Data on the rotational spectrum and structure of $Cl_2O_2$ in the 285-435 GHz range were published in 1989 (Birk *et al.*).

The information listed for each spectral line includes the frequency (MHz) and its estimated or experimental uncertainty, logarithm (base 10) of integrated intensity in units of ($nm^2$ MHz) at 300 K, degrees of freedom in the rotational partition function, lower state energy ($cm^{-1}$) relative to the ground state, upper state degeneracy, species or molecular identifier, and quantum numbers for upper and lower states. The catalogue has an additional, shorter, second file that contains a species directory and additional information for data manipulation.

## ATMOS Catalogue (L. R. Brown)

In order to provide for the needs of, and timely responses to, the Atmospheric Trace Molecule Spectroscopy (ATMOS) mission flown on *SPACELAB 3* (Farmer *et al.* 1986), the investigators adopted both HITRAN and GEISA molecular databases. As the data handling requirements of the ATMOS grew, these databases were expanded to include molecular line parameters of more species in the atmospheric spectra. The ATMOS instrument required data in the wavelength region 580-4500 $cm^{-1}$, but the range and list were expanded to the 1-10,000 $cm^{-1}$ region with over 400,000 entries for 46 species (Brown *et al.* 1987). The database follows the format of AFGL 1982 atmospheric and trace gas compilations (Rothman *et al.* 1982). In most cases the isotopic ratios are those of HITRAN (and GEISA) 1986, and the molecular number code follows HITRAN up to 20. The ATMOS molecular line list is divided into two lists: the main list, which contains more accurate data and which more closely meets the needs of ATMOS, and a supplemental list which contains very preliminary parameters.

## RADEN Data Bank (L. A. Kuznetsova and A. V. Stolyarov)

A data bank has been created at the Department of Chemistry of the Moscow State University by L. A. Kuznetsova and A. V. Stolyarov. This database, known as RADEN, contains information on radiative, magnetic, and energy parameters for approximately 300 diatomic molecules. The compilation contains, in PC-compatible format, data on potential curves, lifetimes of excited levels, dipole moments, electronic transition moments, oscillator strengths, Landé *g*-factors, Franck-Condon factors, and Einstein coefficients. RADEN comprises information derived from more than 2600 publications for about 1100 electronic states and 1200 electronic transitions.

An application program is available to construct potential curves (harmonic, Morse, RKR, and Dunham), to solve the radial Schrödinger equation for both bound and continuum states, to evaluate overlap integrals between vibronic wave functions (Franck-Condon factors, r-centroids, nonradiative transition probabilities), to calculate Einstein coefficients for lines and bands of vibronic and electronic transitions, and to determine electronic transitions moments from experimental data on radiative lifetimes and branching ratios.

RADEN users will receive requested information for particular molecules, states, and transitions, data from original publications (past and future), and reviews or compilations of published data.

## Opacity Calculations (R. L. Kurucz)

R. L. Kurucz, at the Harvard-Smithsonian Center for Astrophysics has collected and calculated atomic and molecular data on energy levels, wavelengths, *gf*-values, and damping constants (Kurucz 1979, 1981, 1992). These data are used for the calculation of statistical line opacities in models of stellar atmospheres. His opacity calculations now include 58 million atomic and molecular lines, for 56 temperatures from 2000 K to 200,000 K, in the range 10 to 10,000 nm and for microturbulent velocities 0,1,2,4, and 8 $kms^{-1}$ and over a range of scaled solar abundances.

Kurucz assembled data for the diatomic molecules $H_2$, CO, and SiO and then, with L. Rossi and J. Dragon, data for the hydrides (CH, NH, OH, MgH, and SiH), CN, $C_2$, TiO, and some of their isotopes. While these molecular data are adequate for statistical opacities at stellar temperatures, Kurucz warns that they are not good enough for detailed spectral comparisons, especially for lines from high vibrational and rotational levels. He will revise all his line lists as improved data are available. He has, for example, CN spectra taken by J. Brault with a Fourier transform spectrometer (FTS) that are substantial improvements over the older data and are directly applicable to identifying weak features in the spectra of the Sun and cool stars. Kurucz now distributes his computed line data on tapes, but he plans to publish the line data on CD-ROMs. He expects to provide tables of all measured data together with his computed energy levels, damping constants, Landé *g*-values, radiative lifetimes, branching ratios, and line *gf*-values.

## Databases for Aeronomics Problems (R. W. Nicholls and M. Nicolet)

Molecular compilations are used by R. W. Nicholls and colleagues at York University, Ontario, and by Marcel Nicolet and colleagues at Institut d'Aéronomie Spatiale de Belgique in support of spectral synthesis programs for research in atmospheric problems. Typically, critically evaluated and compiled, spectroscopic information on molecular line centers and/or structure constants, transition probability data, line profile information, Hönl-London factors, and thermodynamic constants are used for realistic numerical synthesis of molecular spectra. For example, a database to support modelling of the transmission of ultraviolet solar flux in the wavelength region of the photodissociation of molecular oxygen in the mesosphere and stratosphere was assembled and developed. The photodissociation of $O_2$ in the earth's atmosphere depends on the solar irradiance and on the absorption cross section in the wavelength regions corresponding to the $O_2$ Schumann-Runge bands, the $O_2$ Schumann-Runge continuum, and to the $O_2$ Herzberg continuum. The database for Nicolet's theoretical determinations of the rotational structure and absorption cross sections of the $O_2$ Schumann-Runge bands originated from experimental results obtained by the Harvard-Smithsonian Center for Astrophysics (Yoshino *et al.* 1992, 1988, 1987, 1983; Cheung *et al.* 1986a, 1986b 1990) and the Canberra groups (Lewis *et al.* 1986). Nicolet used the experimental cross section measurements and structure of the Schumann-Runge bands at 300 K and 79 K to test his spectral synthesis methods before applying them, for aeronomic purposes, at temperatures between 170 and 270 K.

The results of Nicolet's theoretical modelling have been reported under the general heading of Aeronomic Problems of the Molecular Oxygen Photodissociation, and have been published in a series of papers (Nicolet *et al.* 1986, 1987, 1988a, 1988b, 1989a, 1989b).

R. W. Nicholls and M. W. P. Cann used molecular databases, compiled from their own laboratory work as well as from measurements by the Harvard-Smithsonian Center for Astrophysics and by the Australian groups, to develop extensive facilities for modelling molecular spectral intensities from the microwave to the VUV spectral regions. These data are compiled with the HITRAN format. The results have been published in a series of papers (Nicholls *et al.* 1985, 1988, 1991).

*Acknowledgements:* The author thanks Kelly V. Chance, L. S. Rothman, Peter L. Smith, and K. Yoshino for advice and information on molecular databases. He also thanks Angela Brown for typing the camera-ready manuscript. The author was supported by Harvard College and the Smithsonian Institution.

## References

Berk, M., Friedl, R.R., Cohen, E.A., Pickett, H.M., and Sander, S.P. 1989, *J. Chem. Phys.* **91**, 6588.

Brown, L.R., Farmer, C.B., Rinsland, C.P., and Toth, R.A. 1987, *Appl. Optics* **26**, 5154.

Calfee, R.F. and Benedict, W.S. 1966, *NBS Technical Note 332*, USA Printing Office.

Cann, M.W.P. and Nicholls, R.W. 1991, *Can. J. Phys.* **69**, 1163.

Cheung, A.S.-C., Yoshino, K., Parkinson, W.H., Guberman, S.L., and Freeman, D.E. 1986, *Planet. Space Sci.* **34**, 1007.

Cheung, A.S.-C., Yoshino, K., Parkinson, W.H., and Freeman, D.E. 1986, *J. Molec. Spectrosc.* **119**, 1.

Cheung, A.S.-C., Yoshino, K., Esmond, J.R., Chiu, S.S.-L., Freeman, D.E., and Parkinson, W.H. 1990, *J. Chem. Phys.* **92**, 842.

Conway, R. R. 1988, NRL Memorandum Report 6155.

Coquart, B., Merienne, M.F., and Jenouvrier, A. 1990, *Planet. Space Sci.* **38**, 287.

Farmer, B., Roper, O.F., and O'Callaghan, F.G. 1985, *JPL* publication 87-32.

Farmer, B. and Roper, O.F. 1986, NASA Conference Proceedings CP-2429.

Gallagher, J.W., Brion, C.E., Samson, J.A.R., and Langhof, P.W. 1988, *J. Phys. Chem. Ref. Data* **17**, 9.

Gates, D.M., Calfee, R.F., and Hanson, D.W., and Benedict, W.W. 1964, *NBS Monograph* **71**, USA Printing Office.

Hinteregger, H. E., Fukui, K., and Gilson, B. R. 1981, *Geophys. Res. Lett.* **8**, 1147.

Huber, K.P. and Herzberg, G. 1979, "Molecular Spectra and Molecular Structure IV Constants of Diatomic Molecules", Van Nostrand Reinhold Company, London, Toronto, Melbourne.

Husson, N., Bonnet, B., Scott, N.A., and Chedin, A. 1991, Internal Note, Laboratoire de Metéorologie Dynamique, No. 163.

Jenouvrier, A., Coquart, B., and Merienne-Lafore, M.F. 1986, *Planet. Space Sci.* **34**, 253.

Kirby, K. Constantinides, E. R., Baber, S., Oppenheimer, M. and Victor, G. A. 1979, *Atomic Data and Nuclear Data Tables* **23**, 63.

Kurucz, R.L. 1979, *Ap. J. Suppl. Ser.* **40**, 1.

Kurucz, R.L., 1981, SAO Spec. Rep. No. 390.

Kurucz, R.L., 1991, in "Stellar Atmospheres: Beyond Classical Models", L. Crivellari, I. Hubeny, and D.G. Hummer, ed. *NATO ASI Series*, Kluwer, Dordrecht.

Kyle, T.G. 1968, AFCRL 68-0521.

Lewis, B.R., Berzins, L., and Carver, J.H. 1986, *J. Quant. Spectrosc. Rad. Trans.* **36**, 209.

Lovas, F. J. and Suenram, R. D. 1989, *J. Phys. Chem. Ref. Data* **18**, 1245.

Mathis *et al.* 1983, *Astron. & Astrophys.* **128**, 212.

McClatchey, R.A., Benedict, W.S., Clough, S.A., Burch, D.E., Calfee, R.F., Fox, K., Rothman, L.S., and Garing, J.S. 1973, AFCRL-TR-73-0096 Environmental Paper No. 434.

Nicholls, R.W. 1988, *J. Quant. Spectrosc. Radiat. Transfer.* **40**, 275.

Nicholls, R.W. 1985, *Trans. IAU* **19B**, 146.

Nicholls, R.W. (ed) Series No. 1, *Identification Atlas of Molecular Spectra*, University of Western Ontario (1964/65), York University (1967/72).

Nicolet, M., Cieslik, S., and Kennes, R. 1989a, *Planet. Space Sci.* **37**, 427.

Nicolet, M., Cieslik, S., and Kennes, R. 1989b, *Planet. Space Sci.* **37**, 459.

Nicolet, M., Cieslik, S., and Kennes, R. 1988a, *Planet. Space Sci.* **36**, 1039.

Nicolet, M. and Kennes, R. 1988b, *Planet. Space Sci.* **36**, 1069.

Nicolet, M., Cieslik, S., and Kennes, R. 1987, *Aeronomica Acta*, Brussels, A. no. **318**, 1-340.

Nicolet, M. and Kennes, R. 1986, *Planet. Space Sci.* **34**, 1043.

Park, J.H., Rothman, L.S., Rinsland, C.P., Pickett, H.M., Richardson, D.J., and Namkung, J.S. 1987, "Atlas of Absorption Lines from 0 to 17900 $cm^{-1}$" (HITRAN atlas), NASA Publication 1188.

Pearse, R.W.B. and Gaydon, A.G. 1992, "The Identification of Molecular Spectra", Fifth Edition, Chapman and Hall, London, A. Halsted Press Book, John Wiley & Sons, Inc., New York.

Phillips, J.G. and Davis, S.P. 1963, 1968 Berkeley Analysis of Molecular Spectra, University of Colorado Press, Berkeley.

Pickett, H.M., Cohen, E.A., and Margolis, J.S. 1985, *J. Mol. Spectrosc.* **110**, 186.

Pickett, H.M., Cohen, E.A., Brown, L.R., Rinsland, C.P., Smith, M.A.H., Devi, V.M., Goldman, A., Barbe, A., Carli, B., and Carlotti, M. 1988, *J. Mol. Spectrosc.* **128**, 151.

Pickett, H. M., Poynter, R. L. and Cohen, E. A. Feb. 14, 1992, ***Submillimeter, Millimeter, and Microwave Spectral Line Catalog***, available from National Space Science Data Center, GSFC.

Poynter, L. Robert, and Pickett, Herbert, M. 1984. *J.P.L.*, 80-23, Div 2.

Poynter, L. Robert, and Pickett, Herbert, M. 1985, ***Appl. Optics*** **24**, 2285.

Rosen, B. (ed) 1951, Données spectroscopiues concernant les molécules diatomiques, Herman el Cie Paris.

Rosen, B. (ed) 1970, Données spectroscopiques concernant les molécules diatomiques, 2nd Edition, Pergamon Press, Oxford.

Rothman, L.S., Gamache, R.R., Barbe, A., Goldman, A., Gillis, J.R., Brown, L.R., Toth, R.A., Flaud, J.-M., and Camy-Peyret, C. 1983, *Appl. Optics* **22**, 2247.

Rothman, L.S., Goldman, A., Gillis, J. R., Gamache, R.R., Pickett, H.M., Poynter, R.L., Husson, N., and Chedin, A. 1983, ***Appl. Optics*** **22**, 1616.

Rothman, L.S., Gamache, R.R., Goldman, A., Brown, L.R., Toth, R.A., Pickett, H.M., Poynter, R. L., Flaud, J.-M., Camy-Peyret, C., Barbe, A., Husson, N., Rinsland, C.P., and Smith, M.A.H. 1987, ***Appl. Optics*** **26**, 4058.

Rothman, L.S. 1991, *HITRAN Newsletter* **1**, 1, Nov.

Yoshino, K., Esmond, J.R., Cheung, A.S.-C., Freeman, D.E., and Parkinson, W.H. 1992 *Planet. Space Sci.* in press.

Yoshino, K., Cheung, A.S.-C., Esmond, J.R., Parkinson, W.H., Freeman, D.E., Guberman, S.L., Jenouvrier, A., Coquart, B., and Merienne, M.F. 1988, *Planet. Space Sci.* **36**, 1469.

Yoshino, K., Freeman, D.E., Esmond, J.R., and Parkinson, W.H. 1987, *Planet. Space Sci.* **35**, 1067.

Yoshino, K., Freeman, D.E., Esmond, J.R., and Parkinson, W.H. 1983, *Planet. Space Sci.* **31**, 339.

Printing: Druckhaus Beltz, Hemsbach
Binding: Buchbinderei Schäffer, Grünstadt

# Lecture Notes in Physics

For information about Vols. 1–374
please contact your bookseller or Springer-Verlag

Vol. 375: C. Bartocci, U. Bruzzo, R. Cianci (Eds.), Differential Geometric Methods in Theoretical Physics. Proceedings, 1990. XIX, 401 pages. 1991.

Vol. 376: D. Berényi, G. Hock (Eds.), High-Energy Ion-Atom Collisions. Proceedings, 1990. IX, 364 pages. 1991.

Vol. 377: W. J. Duschl, S.J. Wagner, M. Camenzind (Eds.), Variability of Active Galaxies. Proceedings, 1990. XII, 312 pages. 1991.

Vol. 378: C. Bendjaballah, 0. Hirota, S. Reynaud (Eds.), Quantum Aspects of Optical Communications. Proceedings 1990. VII, 389 pages. 1991.

Vol. 379: J. D. Hennig, W. Lücke, J. Tolar (Eds.), Differential Geometry, Group Representations, and Quantization. XI, 280 pages. 1991.

Vol. 380: I. Tuominen, D. Moss, G. Rüdiger (Eds.), The Sun and Cool Stars: activity, magnetism, dynamos. Proceedings, 1990. X, 530 pages. 1991.

Vol. 381: J. Casas-Vazquez, D. Jou (Eds.), Rheological Modelling: Thermodynamical and Statistical Approaches. Proceedings, 1990. VII, 378 pages. 1991.

Vol. 382: V.V. Dodonov, V. I. Man'ko (Eds.), Group Theoretical Methods in Physics. Proceedings, 1990. XVII, 601 pages. 1991.

Vol. 384: M. D. Smooke (Ed.), Reduced Kinetic Mechanisms and Asymptotic Approximations for Methane-Air Flames. V, 245 pages. 1991.

Vol. 385: A. Treves, G. C. Perola, L. Stella (Eds.), Iron Line Diagnostics in X-Ray Sources. Proceedings, Como, Italy 1990. IX, 312 pages. 1991.

Vol. 386: G. Pétré, A. Sanfeld (Eds.), Capillarity Today. Proceedings, Belgium 1990. XI, 384 pages. 1991.

Vol. 387: Y. Uchida, R. C. Canfield, T. Watanabe, E. Hiei (Eds.), Flare Physics in Solar Activity Maximum 22. Proceedings, 1990. X, 360 pages. 1991.

Vol. 388: D. Gough, J. Toomre (Eds.), Challenges to Theories of the Structure of Moderate-Mass Stars. Proceedings, 1990. VII, 414 pages. 1991.

Vol. 389: J. C. Miller, R. F. Haglund (Eds.), Laser Ablation-Mechanisms and Applications. Proceedings. IX, 362 pages, 1991.

Vol. 390: J. Heidmann, M. J. Klein (Eds.), Bioastronomy - The Search for Extraterrestrial Life. Proceedings, 1990. XVII, 413 pages. 1991.

Vol. 391: A. Zdziarski, M. Sikora (Eds.), Ralativistic Hadrons in Cosmic Compact Objects. Proceedings, 1990. XII, 182 pages. 1991.

Vol. 392: J.-D. Fournier, P.-L. Sulem (Eds.), Large-Scale Structures in Nonlinear Physics. Proceedings. VIII, 353 pages. 1991.

Vol. 393: M. Remoissenet, M.Peyrard (Eds.), Nonlinear Coherent Structures in Physics and Biology. Proceedings. XII, 398 pages. 1991.

Vol. 394: M. R. J. Hoch, R. H. Lemmer (Eds.), Low Temperature Physics. Proceedings. XXX,XXX pages. 1991.

Vol. 395: H. E. Trease, M. J. Fritts, W. P. Crowley (Eds.), Advances in the Free-Lagrange Method. Proceedings, 1990. XI, 327 pages. 1991.

Vol. 396: H. Mitter, H. Gausterer (Eds.), Recent Aspects of Quantum Fields. Proceedings. XIII, 332 pages. 1991.

Vol. 398: T. M. M. Verheggen (Ed.), Numerical Methods for the Simulation of Multi-Phase and Complex Flow. Proceedings, 1990. VI, 153 pages. 1992.

Vol. 399: Z. Švestka, B. V. Jackson, M. E. Machedo (Eds.), Eruptive Solar Flares. Proceedings, 1991. XIV, 409 pages. 1992.

Vol. 400: M. Dienes, M. Month, S. Turner (Eds.), Frontiers of Particle Beams: Intensity Limitations. Proceedings, 1990. IX, 610 pages. 1992.

Vol. 401: U. Heber, C. S. Jeffery (Eds.), The Atmospheres of Early-Type Stars. Proceedings, 1991. XIX, 450 pages. 1992.

Vol. 402: L. Boi, D. Flament, J.-M. Salanskis (Eds.), 1830-1930: A Century of Geometry. VIII, 304 pages. 1992.

Vol. 403: E. Balslev (Ed.), Schrödinger Operators. Proceedings, 1991. VIII, 264 pages. 1992.

Vol. 404: R. Schmidt, H. O. Lutz, R. Dreizler (Eds.), Nuclear Physics Concepts in the Study of Atomic Cluster Physics. Proceedings, 1991. XVIII, 363 pages. 1992.

Vol. 405: W. Hollik, R. Rückl, J. Wess (Eds.), Phenomenological Aspects of Supersymmetry. VII, 329 pages. 1992.

Vol. 406: R. Kayser, T. Schramm, L. Nieser (Eds.), Gravitational Lenses. Proceedings, 1991. XXII, 399 pages. 1992.

Vol. 407: P. L. Smith, W. L. Wiese (Eds.), Atomic and Molecular Data for Space Astronomy. VII, 158 pages. 1992.

Vol. 408: V. J. Martínez, M. Portilla, D. Sàez (Eds.), New Insights into the Universe. Proceedings, 1991. XI, 298 pages. 1992.

# New Series m: Monographs

Vol. m 1: H. Hora, Plasmas at High Temperature and Density. VIII, 442 pages. 1991.

Vol. m 2: P. Busch, P. J. Lahti, P. Mittelstaedt, The Quantum Theory of Measurement. XIII, 165 pages. 1991.

Vol. m 3: A. Heck, J. M. Perdang (Eds.), Applying Fractals in Astronomy. IX, 210 pages. 1991.

Vol. m 4: R. K. Zeytounian, Mécanique des fluides fondamentale. XV, 615 pages, 1991.

Vol. m 5: R. K. Zeytounian, Meteorological Fluid Dynamics. XI, 346 pages. 1991.

Vol. m 6: N. M. J. Woodhouse, Special Relativity. VIII, 86 pages. 1992.

Vol. m 7: G. Morandi, The Role of Topology in Classical and Quantum Physics. XIII, 239 pages. 1992.

Vol. m 8: D. Funaro, Polynomial Approximation of Differential Equations. X, 305 pages. 1992.

Vol. m 9: M. Namiki, Stochastic Quantization. X, 217 pages. 1992.

Vol. m 10: J. Hoppe, Lectures on Integrable Systems. VII, 111 pages. 1992.

Vol. m 11: A. D. Yaghjian, Relativistic Dynamics of a Charged Sphere. XII, 115 pages. 1992.

Vol. m 12: G. Esposito, Quantum Gravity, Quantum Cosmology and Lorentzian Geometries. XVI, 326 pages. 1992.